COLOR SCHEMES

U.S. NAVY

1911 - 1950

Printed in Canada

CURTISS SEAPLANE (A-1) J.M. ELLIOTT

Glenn H. Curtiss and Lt. T.G. Ellyson (the Navy's first pilot) prepare for take-off from Keuka Lake, Hammondsport, New York in June, 1911, during amphibious trials held prior to being delivered to the Navy in July.

COLOR SCHEMES and MARKINGS U.S. NAVY AIRCRAFT 1911 – 1950

by

BILL C. KILGRAIN

illustrations by

BILL C. KILGRAIN

TABLE OF CONTENTS

	Page
Introduction	vii
Acknowledgements	ix
Section 1 - Standard Color Schemes	1
Section 2 - Standard Insignia And Markings	9
Section 3 - Squadron And Unit Coding	20
Photographs A-1 to A-39	31
Section 4 - Squadron Tail Markings	51
Section 5 - Carrier Squadron Colors	62
Section 6 - Carrier Identification Markings	66
Section 7 - Escort Carrier Markings	82
Section 8 - Miscellaneous Markings	93
Photographs B-1 to B-14	103
Section 9 - Special Purpose Aircraft	111
Photographs C-1 to C-20	121
Section 10 - Designation Systems	131
Section 11 - Carrier Aircraft Assignment	144
References	178

INTRODUCTION

This book, which I began writing in 1968, has been planned with the modeler in mind, and it is hoped that this work will help ease the modeler's dilemma as to accurate color schemes and markings for U.S. Navy aircraft.

The text is laid out chronologically within sections that correspond to the finishing steps on a model, i.e. standard finish, standard markings, unit markings, I.D. markings, and special markings. This was done so that a modeler could decide on a period or specific battle to represent, and then finish his model in stages by refering to various sections in the book.

In all illustrations, the basic color scheme of the air--craft is represented by either black (for clarity of markings) or a grey tone. This merely represents the color scheme in use during the period being illustrated and not the actual color of the aircraft. The unit and I.D. markings are shown in color.

The drawings are accurate in all respects except one .... the style of lettering and numbers used are not of the U.S. military type. The style I used, was used for reasons of speed (letra-set decals), conformity of size and shape, and for general neatness of the drawings.

The Carrier Aircraft Assignment tables for World War II cover only a few specific battles because of the amount of space required to go into greater detail during this period. The comments in this section on the role of various units applies to that group or unit, ie. TG 58.4, which it follows unless noted otherwise. This was included, along with in--formation on carrier losses and damage, to add a brief oper--ational history to this section.

Hopefully the modeler will find it less frustrating to complete an accurately finished aircraft with the aid of this book. It is also hoped that this book will find a place on the aviation historian's book shelf because of the markings and photos included in this work.

Bill C. Kilgrain
I.P.M.S./Canada
Victoria, B.C.

19 May, 1973

Below are listed, and explained, some of the letters used throughout this book in conjunction with aircraft carrier names.

CV ..... Heavy (Fleet) Carrier
CVG .... Carrier Air Group

CVB .... Large (Battle) Carrier
CVBG ... Carrier Air Group

CVL .... Light Carrier
CVLG ... Carrier Air Group

CVE .... Escort (Jeep) Carrier
CVEG ... Carrier Air Group

ACV .... early designation for Escort Carriers
VGS .... Escort Scouting Squadron (the letter 'G' was added in the squadron designation of squadrons assigned to Escort Carriers)

The designations CV-9, CVE-27, etc. indicate the carrier class and hull number of each ship concerned. There are class letters for other ships also .... such as BB for Battleship, CH for Heavy Cruiser, and CL for Light Cruisers.

ACKNOWLEDGEMENTS

There are two people who are due a very grateful "Thank You" for their assistance in making this book possible. They are Major J.M. Elliott (U.S.M.C. Ret.) of the Smithsonian Institute, Washington, D.C., and William T. Larkins, author of 'U.S. Navy Aircraft 1921 - 1941'.

Both these men spent many hours patiently doing research for me in answer to my many requests for information. They also made available numerous photos from their collections for inclusion in this book. Most photos received from them have had the photo credit given to the original source of the photograph as per their request.

The photos from Peter M. Bowers' Collection were obtained from Vic Seeley of Castle Graphics, Greenbank, Washington. I also thank Mr. Seeley for his aid in supplying substitute photos for unavailable ones which I had ordered. In most cases his choice of substitute was as good as, or better than, the photos I had ordered.

Thanks also go to the Historian, I.P.M.S./Canada who pro--vided numerous copies of aircraft drawings which aided greatly in preparation of the illustrations in this book.

I also wish to thank my wife for her help in proof reading the manuscript and for her patience during the many evenings of desertion while I was working on this book.

Bill C. Kilgrain

1 STANDARD COLOR SCHEMES

1911 - 1915

During this period there was no 'official' color scheme. Aircraft were usually left in the original factory finish of light buff or off-white; this being the color of the fabric after it had been clear doped. Wood portions, such as struts, were clear varnished and metal portions, such as engine cover plates, were either left natural metal or painted grey. Sea--planes usually appeared brown in color as a result of their fabric being rubberized. Some aircraft had their complete fuselage painted light grey.

1916

SEPTEMBER 20 — Instructions were issued changing the color scheme on N-9 aircraft from grey to opaque yellow varnish.

1917

MARCH 13 — A directive was issued instructing that ALL seaplanes now be painted opaque yellow over--all.

AUGUST 15 — The Curtiss Company was authorized to paint the wings with 'English-Khaki-Grey Enamel' and all manufacturers were to use opaque yellow or clear varnish on floats and hulls.

Many changes to the finish of aircraft were authorized during the next 6 months and it became the tendency to paint aircraft khaki over-all.

1918

MARCH 15 — A directive was issued instructing that ALL new aircraft be painted in low visability naval grey enamel.

During World War I some flying boats were camouflaged with 'Dazzle Paint'. This was a series of stripes and zig-zag lines of a color contrasting to the basic finish of the hull; the pattern used was similar to that used on some surface vessels.

1919

JUNE 21 — Instructions were issued to paint all fabric covered surfaces with aluminum enamel; all other surfaces were to remain painted with naval grey enamel.

1924

FEBRUARY 4 — All squadrons of a station, force or fleet were permitted to uniformly paint the upper wing chrome yellow or other color to increase visi- -bility in case of a forced landing. The use of camouflage or striping on aircraft was to be discontinued.

JUNE 24 — Aircraft were to be painted with aluminum enamel over-all and clear varnish was to be applied to wooden struts and spars. The upper surfaces of the top wing of all aircraft of any station, force or fleet could be painted with naval yellow enamel or other high visibility color.

1925

MAY 29 Naval aircraft were now to be painted aluminum over-all with the upper surfaces of the top wing and horizontal tail painted chrome yellow. Hulls and floats of seaplanes were to be painted navy grey.

About this time, as a ground safety precaution, propeller tips were painted with three 4" bands of red, yellow, and dark blue. Later the rear of the blade, that portion visible to the pilot, was painted black in order to reduce glare for the pilot. When the rear of the blade was painted black, the blue band was usually deleted from the rear surface of the blade. The remaining surfaces of the propeller were painted silver.

1927

A minor change in color scheme came into use about this time and metal surfaces were painted light grey -- almost off-white in shade. Except for this change the color scheme remained as directed in 1925.

1928

A change in the color of hulls and floats on seaplanes occured about this time; the under surfaces, up to the chin, were now painted dark blue or black instead of grey.

1934

By this time aircraft had reverted to the over-all aluminum finish as directed in 1925.

1940

OCTOBER A Basic Non-Camouflage Scheme was authorized.

This consisted of the upper wing surfaces (top wing only on biplanes) being painted glossy chrome yellow, with the yellow extending 5% of the chord onto the under surfaces of the lead--ing edge. The remainder of the aircraft was painted glossy aircraft grey, as were the floats on aircraft so equipped.

DECEMBER 30 — A directive was issued instructing that ALL ship based aircraft now be painted non-specular light grey over-all. Patrol aircraft were to be light grey except for surfaces viewed from above which were to be painted non-specular blue grey. All amphibious aircraft not based aboard ship were to be painted the same as patrol aircraft.

1941

OCTOBER 13 — The Bureau of Aeronautics directed that all fleet aircraft be painted non-specular light grey with all surfaces viewed from above painted non-specular blue grey. Definite demarcation lines were to be avoided by use of a wavy line and then blending the two colors together. On aircraft with folding wings, that portion of the under wing surfaces viewed when folded up was also painted non-specular blue grey.

DECEMBER 26 — It was directed that all shore based aircraft be painted the same as fleet aircraft.

1942

JANUARY 14 — Instructions were issued to use readily remove--able non-specular black paint for purposes of night camouflage. *

* See end of chapter for 'Black Cat' finishes.

1943

FEBRUARY 1 The old two color camouflage scheme was replaced by a new Basic (Non-Specular) Camouflage Scheme. The upper portion of the fuselage was painted non-specular sea blue with non-specular inter--mediate blue applied to the fuselage sides, vertical tail and rudder surfaces. The upper surfaces of the wings and horizontal tail were painted semi-gloss sea blue. All under surfaces were painted non-specular insignia white; this not to extend up the fuselage sides, nor beyond the shadow of the horizontal surfaces. The intermediate blue used on the fuselage sides was blended into the upper and lower colors so as to leave no definite demarcation line. On aircraft with folding wings, that portion of the under wing surfaces viewed when folded up was also painted non-specular intermediate blue. The two colors were again blended into each other so as to leave no definite demarcation line. Leading edges of the wings were painted non-specular sea blue and blended into the upper and lower wing surface colors.

Floats were painted non-specular sea blue on the upper surfaces and non-specular white on the under surfaces, both colors meeting at the chin of the float. Float and wing struts were painted non-specular intermediate blue; on large struts the upper portion was usually painted non-specular white.

Propeller tips were to be painted orange-yellow and the remaining propeller surfaces were to be painted non-specular black. On aircraft where the propellers presented no hazard to personnel the yellow tips were not mandatory. Spinners were to be painted non-specular intermediate blue.

1944

MARCH 22 — The colors of the basic camouflage scheme, as issued in February 1943, were changed from non-specular to glossy finish. Fighter aircraft were to be painted in glossy sea blue over-all. The use of non-specular sea blue in place of glossy sea blue was allowed in areas where it was necessary in order to protect the pilot from sun glare.

A new non-camouflage scheme, aluminum over-all, was authorized for use on aircraft not in a combat area.

OCTOBER 7 — New color schemes were authorized as follows:
Carrier Based Aircraft... glossy sea blue over-all.
Seaplane Transports and Utility Aircraft... glossy sea blue over-all.
Helicopters, Patrol and Observation Aircraft... finished in the basic camouflage scheme as was assigned in February 1943 using non-specular colors.
Landplane Transports and Utility Aircraft... over-all aluminum.

Author's comments:

a) It appears as though aircraft in combat areas were not re-painted in the glossy sea blue finish when it was adopted, instead they retained their original finish throughout their service in combat. The new finish was applied to all aircraft before they left the factory for combat areas.

Anti-submarine Patrol Aircraft (operating in areas where no enemy air opposition was to be expected)...

Scheme I - for use in areas where the weather was mainly clear - All upper surfaces were to be painted dark gull grey, all side surfaces were painted light gull grey, and all under surfaces were painted glossy insignia white. The side surfaces, in the shadow of the wings and horizontal tail, were painted non-specular white. The inside of the cowlings, propeller domes, and propeller blades (out to the inner edge of the cowling opening) were also painted non-specular white.

Scheme II - for use in areas where the weather was mainly overcast - The same as Scheme I except that light gull grey was replaced by non-specular white.

DECEMBER 30 A new color scheme was assigned to landplane patrol aircraft. The upper and lower surfaces of the wings and all horizontal tail surfaces were painted semi-gloss sea blue. The entire fuselage and vertical tail surfaces were painted non-specular sea blue.

1945

MARCH Authorization was given to paint propeller hubs and domes non-specular black; anti-submarine aircraft still retained the use of white on their hubs and domes.

1947

JANUARY 2 New specifications for aircraft finishes were

issued. All shipboard aircraft, water based aircraft, and helicopters were to be painted glossy sea blue over-all. Landplane transports and utility aircraft were to be finished in aluminum over-all.

1950

Apparently, at the outbreak of the Korean War, night fighters were painted non-specular black over-all. Later, when these precautions proved unnecessary, they reverted to the standard finish of glossy sea blue over-all.

W.W.II "Black Cats"

With American strength in the South Pacific increasing steadily during 1942 and 43, the Japanese began to use small vessels, moving at night, to supply their garrisons. The U.S. answer to these tactics was to employ Catalina aircraft equipped with radar to detect and destroy these vessels at night.

December 1942 The first "Black Cat" squadron, VP-12, arrived at Guadalcanal with the first PBYs equipped for night operations. These aircraft were overall black, except for the aircraft number applied in small white figures to both sides of the vertical tail surfaces, and carried no National Insignia.

September 1943 Fleet Air Wing 17 (VP-11, VP-52, VP-101) was organized and began night operations against Japanese shipping in the New Guinea area. These aircraft were finished the same as VP-12 above.

Late 1944 - early 1945 Aircraft of VBP-50, operating in the Philippines, were finished as above with the addition of the National Insignia (white star and bars only) carried as normal.

2 STANDARD INSIGNIA AND MARKINGS

1911 - 1915

During this period the letters 'U.S.N.', with a bar underneath, were carried to distinguish navy aircraft. These letters were placed wherever they would be most prominent, usually on both sides of the vertical tail surfaces and under the starboard wing tip. The letters were normally applied in white, but it appears that dark blue or black was also used.

1916

APRIL 15 A large anchor and a two digit number, both in dark blue, were adopted to designate naval airplanes. The anchor was carried on both sides of the vertical tail surfaces and near the tips of the upper and lower wings. The number was carried inboard of the anchor on the wing tips and also on both sides of the front and rear fuselage.

1917

MAY 19 The first official National Insignia was adopted for use. This insignia consisted of a five pointed white star, with a red center circle, superimposed on a true blue circle. The insignia on the upper wing was carried inboard of the ailerons and had a diameter which fit between the leading and trailing edges of the wing. The insignia on the lower wing was applied in line with that on the upper wing; if the ailerons interfered, then the insignia was moved inboard until the ailerons were cleared. The Insignia was carried in four wing positions.

Stripes consisting of three vertical bars of blue, white, and red were applied to the entire rudder surfaces.

The bureau number was applied to both sides of the rudder, near the top of the white stripe, using black figures 3 inches high. The number applied was preceded by the letter 'A'. The bureau number was also placed on both sides of the rear fuselage in large black figures; if the fuselage color was dark, then white was used instead of black.

1918

FEBRUARY 8 The National Insignia was replaced with roundels of red, blue, and white which were carried in the normal four wing positions. Also at this time the rudder stripes were reversed to red, white, and blue.[1-2-3]

1. This was to conform with standard Allied markings then in use and to aid in identification of American Expeditionary Force aircraft.

2. Some Nieuport 28's used an alternate method of rudder stripes applied as white, blue, and red.

3. The roundels and reversed rudder stripes were, apparently, only applied to those aircraft that were proceeding over--seas. Occasionally an aircraft would be seen with the National Insignia on one wing and a roundel on the other; the reason for this was that all spare wings sent overseas did not have the roundel applied and time was not taken during battle to change the insignia.

1919

AUGUST 19 — A directive was issued instructing that the National Insignia once again be applied to all naval aircraft. At the same time the rudder stripes were reversed to the original blue, white, and red. (The change over was slow and it was almost a year before the last roundel disappeared from naval aircraft.)

1920

The large bureau number carried on the fuselage sides was declared optional. Many aircraft, however, continued to carry the large number until December, 1924, when its removal was ordered.

1924

AUGUST — Authorization was given to apply the wording 'U.S. NAVY', using 8 inch black letters, to both sides of the rear fuselage near the tail.

The bureau number by this time, on many aircraft, had been moved from the white stripe on the rudder and was now carried near the top of the vertical tail surfaces. This was done because the numbers were getting too large for the white stripe and also to allow the size to be slightly increased for better visibility.

DECEMBER — Instructions were issued ordering the removal of the large bureau number from the fuselage sides; this marking had been optional since 1920. (The large number remained in evidence for another year or two on many aircraft as it usually was not removed until the aircraft required repainting.)

1925

The practice of painting the manufacturer's name across the top of the rudder was adopted about this time. In some cases the assigned aircraft name was also painted on the rudder. This marking, though not used by all manufacturers, was used quite extensively by both Curtiss and Vought.

The National Insignia was reduced in size to fit between the leading edge of the wing and the aileron, and moved out--board to a point about mid-aileron. When the insignia was changed the rudder stripes were also reversed to red, white, and blue.

1928

The practice of painting the manufacturer's name on the rudder was discontinued and the naval aircraft designation number was applied in its place. The manufacturer's name and the designation number were carried together until the early 1930's when the manufacturer's name was removed.

1929

The true blue background circle in the National Insignia had, by this time, been changed to insignia blue.[1]

1930

During the latter part of the year, the letter 'A' was

1. This change may have been authorized on a certain date or it may have been gradually changed over the years. I have not, at this time, found any information to clarify this color change.

removed from the bureau number. The letter remained on some aircraft until about 1932 when it was finally painted out, or the aircraft was replaced by a newer type.

1931

JUNE 1 Identification colors and markings were authorized for use by sections within a squadron. The colors and markings are shown below:

Aircraft Number	Section Number and Color	20" wide Fuselage Stripe	Cowling
-1	1st section royal red	yes	complete
-2		no	top half
-3		no	bottom half
-4	2nd section white	yes	complete
-5		no	top half
-6		no	bottom half
-7	3rd section true blue	yes	complete
-8		no	top half
-9		no	bottom half
-10	4th section black	yes	complete
-11		no	top half
-12		no	bottom half
-13	5th section willow green	yes	complete
-14		no	top half
-15		no	bottom half
-16	6th section lemon yellow	yes	complete
-17		no	top half
-18		no	bottom half
-19 and over	Employed as target tugs, instrument trainers, and scouts. No section colors or markings applied but they did carry squadron coding, squadron tail color, and squadron insignia.		

A wing chevron, in the proper section color, was applied to the upper wing surfaces of all aircraft in the squadron. The chevron was normally applied with the point forward, but a few squadrons did use a reversed chevron. Later, when low wing monoplanes came into use, it was necessary to split the chevron and apply half to the wing on each side of the fuse--lage. The purpose of the chevron was to provide the wing men with a visual guide to aid in formation flying.

The squadron code was centralized across the fuselage band, with the —letter— applied on the band itself. The —letter— was applied in black; white was used on bands that were red, blue, or black.

The aircraft-in-squadron number was applied inside the chevron, behind the point, using large black figures. When the chevron was split on low wing monoplanes, the number was applied inside the chevron and near the trailing edge of the wing. In some cases the number was applied outside the chev--ron and near the center of the wing.

Squadrons with two plane sections, based aboard light cruisers, painted the cowlings of BOTH aircraft complete; all other section markings were carried as normal. The section leader's fuselage stripe was applied to aircraft numbered -1, -3, -5, -7, and -9.

Squadrons with four plane sections, based aboard heavy cruisers, painted the first three aircraft as normal. The fourth aircraft had a front to rear stripe painted on the top and bottom of the cowling; the rest of the section markings were applied as normal. The section leader's fuselage stripe was applied to aircraft numbered -1, -5, -9, -13, and -17.

Author's comments:

a) The use of the section leader's fuselage stripe had been, in some squadrons, an unofficial practice since 1925.

b) It appears as though Fighting Three (VF-3), The Diving Eagles, did not paint their cowlings until about 1935.

NOVEMBER — Authorization was given to apply the wording 'U.S. NAVY', in large black letters, to the under surfaces of the lower wing. This was to be applied in a manner enabling it to be read by an observer as the aircraft approached.

1932

JANUARY — The wording 'U.S. NAVY' was ordered removed from the under surfaces of the lower wing.[1]

1940

MARCH 19 — A directive was issued instructing that the National Insignia was to be applied to both sides of the forward fuselage on all aircraft participating in the Neutrality Patrol. The insignia was to have a diameter of 24 inches on carrier based aircraft and a diameter of 36 inches on larger patrol aircraft.

1941

FEBRUARY 26 — Instructions were issued which discontinued the use of all section markings and colored unit tail markings.

It was also directed that the National Insignia be removed from the upper right and the lower left wing surfaces. The distance from the wing

1. This short lived marking remained in evidence for a few more months until it was finally painted out on all air--craft.

tip to the center of the insignia was to be equal to the chord length of the wing. On float equipped planes the insignia was to be applied inside the floats and as near to this location as possible. The diameter of the insignia was to be the distance from the leading edge to the aileron, but was not to exceed 60 inches. The National Insignia was also to be applied to both sides of the rear fuselage of all aircraft and to both sides of the bow on all flying boats. The diameter of the fuselage insignia was not to exceed 24 inches.

The wording 'U.S. NAVY' was reduced to 1 inch high black letters, moved from the rear fuselage to the vertical tail, and applied as 'NAVY' above the bureau number.

The use of vertical red, white, and blue rudder stripes became mandatory.

1942

JANUARY 5 A change in regulations authorized the National Insignia to once again be carried in four wing positions. The insignia was to be carried inboard from the tip a distance equal to 1/3 of the distance from the tip to the fuselage, and was to be as large as possible without extending onto the aileron. The insignia on the fuselage was to be 24 inches in diameter; if this was too large to be applied the insignia was to be moved forward on the fuselage as necessary.

Stripes, consisting of seven red and six white horizontal bars, were applied to the entire rudder surfaces. The balanced portions of the rudder which extended past the hinge line were painted the same color as the tail surfaces.

The National Insignia and the rudder stripes were applied using non-specular colors.

MAY 15 The red disc in the center of the National In- -signia was ordered removed. The use of rudder stripes was also to be discontinued.

During 'Operation Torch', October - November, all aircraft participating had a yellow surround applied to the National Insignia on the fuselage and the under wing surfaces. The yellow surround was applied by hand and as a result usually had a very ragged edge.

1943

FEBRUARY 1 The National Insignia was ordered removed from the upper right and the lower left wing surfaces. The wing insignia were also decreased in size so that the diameter was equal to 75% of the distance from the leading edge to the aileron and was not to be less than 30 inches.

JUNE 28 A directive was issued which added white rec- -tangles on the left and right sides of the blue background circle in the National Insignia. The whole insignia was also given an insignia red outline border.

The practice was adopted about this time of using a light grey for the white portions of the National Insignia on the upper wing surfaces. This grey color consisted of one part insignia white and one part light grey.

SEPTEMBER 4 The red outline of the National Insignia was changed to insignia blue. Wing insignias were to have a maximum diameter of 60 inches and a minimum of 30 inches. Fuselage insignias were to be no more than 50 inches and no less than 20 inches in diameter. Night fighters were to carry insignias having a 25 inch diameter in all locations.

1946

JUNE — It was authorized that the blue background circle and blue border of the National Insignia could be omitted when applied to aircraft that were finished in sea blue or black over-all.[1]

1947

JANUARY 14 — A red horizontal stripe, centered on the white rectangles, was added to the National Insignia.

1948

During the year a change in location of the bureau number, designation number, and service markings occured. The naval aircraft designation number was moved from the rudder and now applied below the leading edge of the horizontal tail. The wording 'NAVY' and the bureau number were removed from the ver--tical tail and centered under the aircraft designation number. The designation number and 'NAVY' were increased in size to 2 inches, while the bureau number was increased to 4 inches in height. White lettering was used on dark surfaces and black was used on light colored surfaces.

1. There also appears to have been unofficial use of this type of insignia on some fighter aircraft during the latter part of World War II.

1950

FEBRUARY

High visibility markings were adopted for use on all naval aircraft. The word 'NAVY' was removed from the grouping under the tail, increased to a minumum height of 12 inches, and re-located immediately behind the National Insignia on the rear fuselage. This was to be white on aircraft with a dark color scheme and black on a light colored aircraft.

The National Insignia on many jet aircraft, by this time, was applied to both sides of the nose instead of the rear fuse--lage. These aircraft carried 'NAVY' as normal on both sides of the rear fuselage.

All lettering on night fighters, at the outbreak of the Korean War, was to be non-specular red. Later, when these pre--cautions proved to be unnecessary, all lettering was repainted in white as normal.

3 SQUADRON AND UNIT CODING

1918

Flying boats operating along the coast of Great Britain, Ireland, and France carried a large white letter and number on both sides of the forward fuselage. This indicated the aircraft's home base and also the individual station number assigned to the aircraft. These aircraft did not carry the large bureau number on the fuselage sides as was standard.[1] Two examples of this marking are shown below:

K 11 ... Curtiss H-16, aircraft number 11, from Killingholme, Lincolnshire, England.

T 7 .... Curtiss HS-1, aircraft number 7, from Treguier, France.

1919

Naval aviation was still very small and few squadron size units were possible. Aircraft attached to the Air Det--achment, Pacific Fleet, apparently carried a letter - number combination on each side of the fuselage, e.g. 'P-10'.[2]

1923

In December a system of unit markings was adopted for use and consisted of a number-letter-number group. The first number was that of the squadron, the letter indicated the

1. Some aircraft used black letters and numbers with a white outline.
2. This probably indicated the role and position of the air--craft within the Air Detachment.

squadron classification, and the last number identified the aircraft within the squadron. The group characters were as large as possible, with the first figure located in line with the cockpit, and were carried on both sides of the fuselage. The unit markings were also applied to the top of the upper wing and under the lower left wing. The fuselage marking, on flying boats, was carried on both sides of the hull for--ward of the wings.

The letters assigned to each squadron class are indicated as shown below:

F - Combat or Fighting Squadrons
T - Torpedo and Bombing Squadrons
S - Seaplane Patrol Squadrons
O - Scouting, Spotting, and Battleship Observation Squadrons

The letter 'O' was replaced by a slash mark or a dash, in the squadron markings, to prevent confusion with the number zero. Examples of these markings are shown below:

1-T-6	sixth plane, Torpedo and Bombing Squadron One (VT-1)
6-1	first plane, Observation Squadron Six (VO-6)
6/2	second plane, Observation Squadron Six

Aircraft that were assigned to a ship or station, but were not part of a squadron, carried the name of the ship or station on both sides of the fuselage. The name was followed by the number of the aircraft on the ship or station. Shown below are two examples:

U.S.S. LANGLEY 6	sixth plane assigned to a ship
N.A.S. ANACOSTIA	plane assigned to a Naval Air Stn.

Aircraft assigned to a squadron based at a Naval Air Sta--tion carried two letters abbreviating the name of the station. If the station name was one word, the first and last letters were used; if the name was two words, the first letter of each word was used. These letters were followed by a dash and the number of the aircraft within the squadron. Where more than one squadron operated from the same station, the squadron number was inserted in front of the letters representing the station. Examples of these markings are shown below:

PA-7 seventh plane, Pensacola
HR-5 fifth plane, Hampton Roads
2HR-9 ninth plane, second squadron, Hampton Roads

In some cases a squadron would be assigned a specific role. Where this occured the classification letter would be inserted between the station abbreviation and the aircraft number. This marking would be carried as shown below:

6-GL-S-3 third plane, Patrol Squadron Six, Great Lakes

Aircraft belonging to ship board squadrons carried the squadron marking as normal on the fuselage sides. In many cases the name of the ship was also carried below the cockpit.

1924

Squadron markings, as applied in 1923, proved to be un-satisfactory; the markings were now reduced to both sides of the fuselage only. Some aircraft did, however, continue to carry the unit markings above the upper wing until the late 1920's.

1925

Early in the year a new unit marking, the letter 'J'... indicating Transportation Squadrons, began to appear. This letter was assigned to Utility Squadrons in October.

1927

During the year a new system of unit markings was adopted

for use by Utility Squadrons which were assigned to Naval Dis--tricts and were not operating with the fleet. The district number was included in the marking and carried until the early 1930's. An example of this marking is shown below:

VJ-D11-5 fifth plane, Utility Squadron, 11th Naval District

JULY 1 The letter 'B', indicating Battle Force, and the letter 'S', indicating Scouting Force, were added to the squadron designation. These letters were NOT applied to the aircraft and were used for administration purposes only. Torpedo and Utility Squadrons attached to the Scouting Force did, for a short time, apply a small 's' to the marking on their aircraft. The aircraft of Torpedo Squadron 5 (VT-5) with the Asiatic Fleet also used an unofficial application of the letter 'A' in their unit marking. Examples of these markings are shown below:

9-Ts-4	fourth plane, Torpedo Squadron Nine (VT-9), Scouting Force
1-VT-5A	first plane, Torpedo Squadron Five (VT-5), Asiatic Fleet

About this time the Navy formed Bombing Squadrons and the letter 'B' was assigned as their squadron classification letter. The letter 'T', which had formerly designated Torpedo and Bomb--ing Squadrons, was retained by the now seperate Torpedo Squad--rons.

1929

Early in the year large patrol planes were formed into Patrol Squadrons and assigned the letter 'P' as their squadron classification letter. Smaller scout planes were formed into Scouting Squadrons and the letter 'S', formerly for Seaplane

Patrol Squadrons, was assigned as their unit classification letter.

1930

During the later part of the year the dash or slash marking used by Observation Squadrons was dropped and the letter 'O' adopted for use in its place.

1937

JULY 1 — Fleet Squadron Re-organization occured.

Carrier squadrons were re-numbered to conform to the hull number of the carrier they were based aboard, e.g. Fighting Six (VF-6B) aboard the U.S.S SARATOGA (CV-3) was re-numbered Fighting Three (VF-3).

Cruiser Scouting Squadrons were re-numbered to conform with the number of the cruiser division to which they were attached. This was the same numbering system used by Battleship Observation Squadrons.

The use of suffix letters to indicate fleet assignment was discontinued, except for Naval Districts.

The markings of Cruiser Scouting Squadrons were changed slightly at this time. The letter 'C' was added to their squadron designation and the coding on their aircraft was marked as 'CS'; the old 'S' was still retained for use by Scouting Squadrons based aboard carriers.

FLEET SQUADRON RE-ORGANIZATION -- July 1st

OLD SQDN. NUMBER	NEW SQDN. NUMBER	REMARKS
VB-5B	VB-2	remained aboard Lexington (CV-2)
VF-2B	VF-2	" " " " " "
VS-3B	VS-2	" " " " " "
VB-1B	VT-2	moved from Ranger to Lexington
VB-2B	VB-3	remained aboard Saratoga (CV-3)
VF-6B	VF-3	" " " " " "
VS-2B	VS-3	" " " " " "
VT-2B	VT-3	" " " " " "
VB-3B	VB-4	moved from Lexington to Ranger
VF-5B	VF-4	remained aboard Ranger (CV-4)
VS-1B	VS-41	" " " " " "
VS-4B	VS-42	moved from Saratoga to Ranger
VB-7B	VB-5	remained aboard Yorktown (CV-5)
VF-7B	VF-5	" " " " " "
VS-7B	VS-5	" " " " " "
VT-7B	VT-5	" " " " " "
VB-8B	VB-6	remained aboard Enterprise (CV-6)
VF-8B	VF-6	" " " " " "
VS-8B	VS-6	" " " " " "
VT-8B	VT-6	" " " " " "
VS-5B	VCS-2	
VS-6B	VCS-3	
VS-9S	VCS-4	
VS-10S	VCS-5	
VS-11S	VCS-6	
VS-12S	VCS-7	
VS-14S	VCS-8	

VO Squadrons remained as they were, dropping the 'B' suffix, and being listed as VO-1, -2, -3, and -4.

1939

JULY 1 Patrol Squadrons were re-numbered according to the wing to which they were attached, the first number indicated the wing and the second number was the squadron within the wing. Thus VP-26 indicated the sixth squadron attached to Patrol Wing Two.

PATROL SQUADRON RE-ORGANIZATION -- July 1st

OLD NUMBER	NEW NUMBER	BASE AND AREA OF OPERATIONS
VP-7	VP-11	NAS San Diego; West Coast
VP-9	VP-12	" " " " "
---- 1	VP-13	" " " " "
VP-21	VP-14	" " " " "
VP-1	VP-21	NAS Pearl Harbor; Hawaii
VP-4	VP-22	" " " "
VP-6	VP-23	" " " "
VP-8	VP-24	" " " "
VP-10	VP-25	Midway
VP-18	VP-26	
VP-2	VP-31	FAB Coco Solo, Canal Zone
VP-3	VP-32	" " " " "
VP-5	VP-33	" " " " "
VP-16	VP-41	NAS Seattle; North West Coast
VP-17	VP-42	" " " " "
VP-19	VP-43	" " " " "
VP-20	VP-44	" " " " "
----- 1	VP-45	FAB Sitka; Alaska
VP-12	VP-51	NAS Norfolk; East Coast
VP-14	VP-52	" " " "
VP-15	VP-53	" " " "
VP-11	VP-54	" " " "

1. New units -- operated late 1939 only.

1941

The squadron number, mission code, and aircraft number were painted in white figures,12 inches high, ahead of the National Insignia on both sides of the fuselage; on patrol planes it was applied behind the bow insignia. The aircraft number was also applied to both sides of the cowling using white figures 4 inches high.

During the latter part of the year the aircraft number, on some aircraft, was also carried on the leading edge of the wings.

DECEMBER — The color of the squadron mission code was changed to non-specular black.

1942

The pre-war style of mission coding was soon changed, in combat zones, to just the squadron mission letter and the air--craft number.

MARCH 9 — The first of 13 Transport Squadrons was formed and the letter 'R' was assigned as their mission code letter.

During the summer the aircraft number began to be applied to the front of undercarriage doors.

1943

JANUARY 12 — The Chief of Naval Air Operational Training issued instructions that all aircraft operating from bases under his command would be marked with a letter-letter-number group for identifi--cation. The first letter identified the base, the second letter indicated the squadron type, and the number represented the aircraft within the unit. When more than one unit operated from the same base, a number was added behind the station letter to identify each additional unit. Thus J2-F-18 would identify the eighteenth air--craft of Operational Training Unit #2, Fighter Training Unit, based at Jacksonville.

By late summer only the aircraft number was carried on the fuselage, and on many aircraft the number had been moved to the vertical tail surfaces. The number '00' was reserved for Air Group Commanders.

1946

During the year VB and VT squadrons were replaced by VA (Attack) squadrons, whose primary mission was attacking surface targets.

NOVEMBER 7 Identification letters, 36 inches high, were assigned to all carriers, and to all wings, groups and squadrons not assigned to carrier duty. These letters were applied in white on dark colored aircraft, and in black on light colored aircraft.

Fleet Air Wings carried a two letter figure in which the first letter identified the wing and the second letter indicated the squadron within the wing. The squadron letters were assigned alphabetically to each unit in the wing. The letters assigned to the wings are shown below:

1st Fairwing (Marianas)........... A
2nd Fairwing (Hawaii)............. B
3rd Fairwing (Canal Zone)......... C
4th Fairwing (Washington)......... D
5th Fairwing (Virginia)........... E
7th Fairwing .... H
10th Fairwing (Philippine Islands). L

11th Fairwing .... M
14th Fairwing (California)....... S
18th Fairwing .... W

NOVEMBER 15 — Carrier squadrons were limited to VF and VA, thus eliminating the VBF, VB, and VT units.

Utility Squadrons changed from VJ to VU for the unit designation.

DECEMBER 12 — The letters previously assigned to carriers were discontinued and the letters were now assigned to Carrier Air Groups instead.

1947

JANUARY 2 — Unit identification letters, issued in November were ordered displayed on both sides of the vertical tail and on the upper right and lower left wing tips.

JANUARY — A new system of numbering aircraft within a unit was adopted for use. Aircraft were now numbered as shown below:

Carrier based aircraft....
- 100 - Air Group Commander
- 101 and up - first fighter squadron
- 201 and up - second fighter squadron
- 301 and up - first attack squadron
- 401 and up - second attack squadron
- 501 and up - third attack squadron

Shore based aircraft......
the last three digits of the bureau number

These numbers were carried on the forward fuse--lage in large white or black numbers depending on the basic finish of the aircraft.

1949

Many jet aircraft, by this time, were carrying the air-number ahead of the National Insignia on the nose of the aircraft.

1950

FEBRUARY

Authorization was given to apply the full squadron designation under the word 'NAVY', in 8 inch high white letters, on both sides of the rear fuselage. The letter 'V', formerly used for administration only, was also included in the designation on the aircraft.

Carrier squadrons were assigned two or three digit numbers, the first number being the same as the parent air group. Thus VF-51 indicated the first squadron attached to Air Group 5 aboard the U.S.S. VALLEY FORGE.

The aircraft number, by this time, was also carried in-board of the ID letters on the upper right and lower left wing tips. The aircraft number was normally about 3/4 of the height of the ID letters and applied in the same color.

CURTISS TRIAD (A-1) NAVY DEPT., NATIONAL ARCHIVES (80-G-650864)

The Navy's first airplane ready for catapult tests at Annapolis on July 31, 1912. Notice the white 'U.S.N.' on the tail, this was the first marking adopted to identify Navy aircraft.

CURTISS N-9 PETER M. BOWERS' COLLECTION

These aircraft carry the dark blue anchor used to indicate naval airplanes from mid-April, 1916, to mid-May, 1917. The aircraft number on the forward and rear fuselage is also dark blue and was a standard marking during this period.

The N-9 was a seaplane version of the Army's JN-4B having increased wingspan and an improved dual-ignition, 100 HP, engine. The N-9 was the Navy's standard trainer during World War I and the early 1920's.

CURTISS HS-1 NAVY DEPT., NATIONAL ARCHIVES (80-G-72979)

Flying boats based at N.A.S. Trequier, France, in 1918. These aircraft were used for patrol over the English Channel. Notice that these aircraft do not carry the large fuselage bureau number as was normal at this time. Aircraft which operated along the coast of Great Britain, Ireland, and France carried a large white letter and number on both sides of the forward fuselage. This indicated the aircraft's home base and also the individual station number assigned to the aircraft.

NAVAL AIRCRAFT FACTORY F-5L (3665) PETER M. BOWERS' COLLECTION

The fuselage striping on this aircraft, circa 1922, may have been applied for added visi--bility during fleet maneuvers. Marine Seaplane Patrol Squadron One (VS-1M) had a similar striped aircraft in 1922, except that the vertical tail was not striped.

Another F-5L, circa 1921, had the fuselage painted with a large yellow and black checker--board pattern for added visibility during fleet maneuvers.

Author's comment: The color of the stripes is unknown, but to me they appear to be red and white. I make this assumption because the color of the dark stripes appears to be the same as the trailing rudder stripe, which should be red; the light stripes, to me, are the same color as the white stripe on the rudder and the white aircraft number, the first digit is '6', which is partially visible ahead of the wing tip float.

MARTIN M2O-1 (A-6453) U.S. NAVY (1218)

Three-place observation plane, early 1923. Note rear cockpit with machine gun ring mount. The M2O was also built by the Naval Aircraft Factory as the NO-1.

OUGHT UO-1 (A-6607) PETER M. BOWERS' COLLECTION

ne of two aircraft forming the Ship Air Unit aboard the U.S.S. TENNESSEE in late 1924. The -1 was the standard battleship and cruiser based observation aircraft until replaced by the ought O2U-1 in 1928.

CURTISS F6C-1 (A-6969) PETER M. BOWERS' COLLECTION

Squadron Leader's aircraft of Fighting Two (VF-2) based aboard the U.S.S. LANGLEY (CV-1) in 1926. The aircraft still carries the unit markings on the upper wing; this location had been discontinued in 1924 but remained on some aircraft until the late 1920's. Notice also the unusual placement of the section color, in the form of a disc, on the upper wing. The tail and fuselage band are red, as is the disc on the upper wing.

VOUGHT VE-9 (A-6478) PETER M. BOWERS' COLLECTION

Third plane of the 11th Naval District's Utility Squadron, 1927. This type of unit marking was used by Utility Squadrons which were assigned to Naval Districts and were not operating with the fleet. The district number was included in the marking and carried until the early 1930's.

CURTISS F6C-4 (A-7403, A-7404) PETER M. BOWERS' COLLECTION

Aircraft that were assigned to a Naval Air Station, but were not part of a squadron, carried the name of the station on both sides of the fuselage. The above aircraft based at Anacostia in 1928 are Curtiss F6C-3s that were converted from liquid to air cooled engines.

MARTIN T3M-2 PETER M. BOWERS' COLLECTION

Ninth aircraft of Torpedo Squadron Nine (VT-9S), attached to the seaplane tender U.S.S. WRIGHT, making a torpedo drop in early 1928. The small letter 's' indicated Scouting Fleet and was not normally applied to the aircraft; this letter was used for administration pur--poses, as was 'B' for Battle Fleet and 'A' for Asiatic Fleet.

VOUGHT O2U-1 CORSAIR (A-7912, A-7911) NAVY DEPT., NATIONAL ARCHIVES (80-G-426930

Aircraft of Battleship Observation Squadron Two (VO-2) operating from the U.S.S. FLORIDA in 1929. The slash mark was used to prevent confusion with the number zero, which could have risen from the use of their assigned class letter 'O'. During 1930 the slash marking was discontinued and the class letter 'O' carried instead.

CURTISS F6C-3 (A-7144) PETER M. BOWERS' COLLECTIO

Aircraft number 23 from squadron based at Hampton Roads, Virginia, in 1930. Aircraft assign to a squadron based at a Naval Air Station carried two letters abbreviating the name of the station. If the station name was one word, the first and last letters were used: if the nam was two words, the first letter of each word was used. Where more than one squadron operate from the same station, a squadron number was inserted in front of station letters.

MARTIN PM-1 (8296) NAVY DEPT., NATIONAL ARCHIVES (80-G-180228)

Third aircraft of the first section from Patrol Squadron Eight (VP-8S), 1931. Note the under wing bomb racks and the squadron coding applied to the upper right wing. VP-8S replaced its PM-1's with Hall-Aluminum PH-1's in 1932. The PM-1 was a five-place open-cockpit all metal flying boat with a top speed of 117.5 miles per hour. Notice also the 'Excellence "E"' and the squadron insignia in front of the squadron marking on the forward fuselage.

MARTIN T4M-1 (7893) U.S. NAVY (189837)

Third aircraft of the fifth section, Torpedo Squadron Two (VT-2), circles the U.S.S. SARATOGA (CV-3) for a landing on 7 March 1932. Note the squadron insignia just above the wing root and the ship's name SARA on the rear of the flight deck. Also visible is the chevron and aircraft number above the upper wing. Tail color probably green.

CONSOLIDATED P2Y-1 NAVY DEPT., NATIONAL ARCHIVES (80-G-426937)

The first two sections of Patrol Squadron Ten (VP-10F) in formation during late 1933. Three seaplane records were broken on January 10 - 11, 1934, when VP-10F flew six P2Y-1's in non--stop formation from San Francisco to Pearl Harbor, Hawaii. The aircraft of VP-10F carried normal section colors and markings plus all aircraft in squadron carried black tail stripes.

The P2Y was a five-place patrol bomber with a gross weight of approximately 20,500 lbs. and had a top speed of 138.5 miles per hour.

VOUGHT O2U-3 (8229) PETER M. BOWERS' COLLECTION

Third aircraft, third section, of Cruiser Scouting Squadron Nine (VS-9S) operating with Cruiser Division Four in 1934. VS-9S operated four plane sections due to the Cruiser Division being composed of heavy cruisers. Tail stripe is white.

BOEING F4B-4 (9018) PETER M. BOWERS' COLLECTION

Green-tailed third aircraft, third section, of Fighting Three (VF-3B) 'The Striking Eagles' parked on an outrigger boom aboard the U.S.S. RANGER (CV-4) in August, 1934. The outrigger booms solved the problem of temporary storage during deck handling of aircraft.

CONSOLIDATED P2Y-3 PETER M. BOWERS' COLLECTION

Squadron Commander's aircraft of Patrol Squadron Seven (VP-7F) attached to the seaplane tender U.S.S. WRIGHT in 1936. Tail color is solid blue. In 1937 the tail markings were changed to double horizontal blue stripes.

GRUMMAN F2F-1 (9652, 9675, 9692) U.S. NAVY

The second section, Fighting Squadron Two (VF-2B) based aboard the U.S.S. LEXINGTON (CV-2) flying in line abreast formation, early 1937. The section markings adopted in 1931 are clearly shown in the photograph.

The section leader's aircraft 2-F-4 has the fuselage band and complete cowling painted white. 2-F-5 has the upper half of the cowling painted, and 2-F-6 has the lower half of the cowling painted.

Tail color is solid lemon yellow.

VF-2 was the only carrier squadron that did not change number during squadron re--organization which took place on July 1st. VF-2 remained on the LEXINGTON and retained yellow tail.

CONSOLIDATED PBY-2 CATALINA PETER M. BOWERS' COLLECTION

Third aircraft, fourth section, of Patrol Squadron Eleven (VP-11F) at the factory ready for delivery on May 15th, 1937. VP-11 was formed in late 1936 and was the first unit to receive the new PBY-2 models. Tail stripes are green. The large waist gun blisters,seo familiar dur--ing World War II, did not appear until 1939.

CONSOLIDATED PBY-1 PETER M. BOWERS' COLLECTION

Third aircraft, first section, of Patrol Squadron Twelve (VP-12) attached to the seaplane tender U.S.S. LANGLEY in late 1937. The tail stripes are yellow. Both VP-11 and VP-12 were attached to the LANGLEY when she was converted from a carrier to a seaplane tender in 1937. Note the underwing bomb racks and the squadron crest just behind the wing struts.

CURTISS SOC-1 SEAGULL (9858) PETER M. BOWERS' COLLECTION

Third section leader's aircraft of Cruiser Scouting Squadron Three (VCS-3) based aboard the U.S.S. CINCINNATI in 1937. Battleship and Cruiser based aircraft often changed to wheels for temporary shore station operation. Tail stripe is red. Note the 'E' in machine gunnery, the hash marks under the small 'M' indicate that the pilot has re-qualified four years in a row. Note also the 'winged seahorse' squadron insignia. Note 'CS' marking adopted in July 1937.

CURTISS SBC-3 (0533) SMITHSONIAN INSTITUTE (A-452)

Fourth section leader's aircraft, Scouting Squadron Five (VS-5) based aboard U.S.S. YORKTOWN and photographed at a shore station on September 29th, 1937. After July 1st the letter 'S' indicated Scouting Squadrons that were not assigned to Cruisers -- these units being identified by the letters 'CS'. 5-S-10 has a black fuselage band and cowling with the tail painted solid red.

CURTISS SOC-1 (9877) SMITHSONIAN INSTITUTE (A-53266)

Utility aircraft number 2, Patrol Wing Five, attached to the seaplane tender U.S.S. PATOKA in early 1940. Note the Neutrality Patrol Star on the forward fuselage; this marking was authorized in March for use on all aircraft operating with the newly formed Neutrality Patrol.

GRUMMAN F3F-1 (0262) U.S. NAVY (454477)

Third section leader's aircraft, Fighting Seven (VF-7), USS WASP (CV-7), mid-1940. Note the newly designed Blue Burgler Wasp squadron insignia and the Neutrality Patrol Star on the forward fuselage. Note also the pilot's name and rank, Lt.(jg) F.K. SLASON, on the bottom frame of the canopy. Tail color is black. Fuselage band and cowling are blue.

GRUMMAN J2F-4 DUCK (1651) PETER M. BOWERS' COLLECTION

Aircraft based at N.A.S. Alameda in 1941. Aircraft is finished in the Basic Non-Camouflage color scheme, glossy aircraft grey with glossy chrome yellow upper wing surface, as introduced in October 1940.

DOUGLAS SBD-3 DAUNTLESS SMITHSONIAN INSTITUTE (A-48723J)

Aircraft at El Segundo Plant ready for delivery on 4 March, 1942. Color scheme is blue--grey upper surfaces with light grey under surfaces as adopted in October 1941; the large National Insignias were adopted for use in January 1942. The red center of the National Insignia and the rudder stripes were removed on May 15th, 1942.

GRUMMAN F4F-4 WILDCAT NAVY DEPT., NATIONAL ARCHIVES (80-G-30311)

An aircraft of Fighting Forty-One (VF-41) takes off from the U.S.S. Ranger (CV-4), during Operation Torch, October - November 1942. The yellow ring around the fuselage insignia was carried by all U.S. Navy aircraft participating in the operations against French North Africa.

GRUMMAN F4F WILDCATS J.M. ELLIOTT

Aircraft flying formation in August, 1943. The white rectangles were added to the National Insignia in late June, as was the red outline of the insignia. In September the red outline of the insignia was changed to insignia blue. The aircraft are finished in the three color Basic (Non-Specular) Camouflage Scheme as adopted in February, 1943.

GENERAL MOTORS FM-1 WILDCAT NAVY DEPT., NATIONAL ARCHIVES (80-G-383829)

An aircraft of Composite Squadron Fifty-Eight (VC-58) almost over the side of the U.S.S. Block Island (CVE-21), in May 1944. The aircraft is finished in dark gull grey and white scheme adopted for use in the Atlantic Theater in early 1944; in October this scheme was designated as ASW Scheme II.

EASTERN TBM-1C AVENGER

PHIL YANT

Aircraft of Torpedo Squadron One (VT-1) about to be launched from the U.S.S. YORKTOWN (CV-10) during the Battle of the Philippine Sea in June, 1944. The YORKTOWN operated with Task Group 58.1 of Task Force 58 during this operation. Notice YORKTOWN's white stripe on the tail.

CURTISS SB2C-3 HELLDIVER

U.S. NAVY (304753)

Aircraft of Bombing Squadron Three (VB-3) from the U.S.S. YORKTOWN (CV-10) covering the land--ings on Iwo Jima during D-Day, 19 February 1945. During this operation the YORKTOWN was with Task Group 58.4 of Task Force 58. Note the new marking assigned to the YORKTOWN in October 1944. The marking was also carried on the upper right wing as can be seen on #110 and 107.

CHANCE VOUGHT F4U-1D CORSAIR PHIL YANT

Aircraft of Fighting Five (VF-5) preparing for take-off from the U.S.S. FRANKLIN (CV-13), with Task Group 58.2, to provide Combat Air Patrol for Task Force 58 during the capture of Iwo Jima in March, 1945. Notice the tape over the gun ports and number on landing gear door.

CONSOLIDATED PBY-5A (48318) U.S. NAVY (700504)

Aircraft of Patrol Squadron Sixty-Three (VP-63), 14 August 1945, carrying underwing retro--rockets for anti-submarine warfare. After the capture of Morocco in December, 1942, part of VP-63 was based at Port Lyautey and participated in the Gibraltar Straits "barrier patrol". Aircraft is finished in ASW Scheme II adopted in October 1944. Note the penquin (?) figure just behind the national insignia on the forward fuselage.

CURTISS SB2C-4E U.S. NAVY (278819)

Aircraft from U.S.S. SHANGRI LA (CV-38) on photo mission, 17 August 1945. The carrier marking were changed from 'G' symbols to single or double letters on July 27th, 1945. Notice the new color scheme of glossy sea blue over-all which was adopted for use on all carrier based air--craft in October 1944. Aircraft pilot is Ens. M.L. Skinner and the photographer is E. Tuohey The aircraft is probably from VB-85, as Air Group 85 was aboard the Shangri-La at this time.

GRUMMAN F6F-5 HELLCAT NAVY DEPT., NATIONAL ARCHIVES (80-G-344442)

Aircraft of Fighting Forty (VF-40) from the U.S.S. SUWANEE (CVE-27) taking off for a surveil--lance flight over the Japanese home islands, 30 August 1945. Markings were assigned to CVEs in June 1945.

GRUMMAN F8F-1 (95320) PETER M. BOWERS' COLLECTION

Commander's aircraft, Air Group Nineteen, in June, 1947. The letter 'B' was assigned to the U.S.S. BOXER in November, 1946, under the new carrier marking system; in December, 1946, the letters were re-assigned to the air groups rather than the carriers. Note the number 100, reserved for Air Group Commanders, and the unusual extention of the National Insignia with the air group number on it.

GRUMMAN F8F-2 BEARCAT (121703) PETER M. BOWERS' COLLECTION

Aircraft is shown in color scheme and markings appropriate for shore based aircraft during the period 1948 - February 1950. The aircraft number 703 is the last three digits of the bureau number, carried by all shore based aircraft in this location. Note the new location of the bureau number, designation, and service below the horizontal stabilizer.

McDONNELL F2H BANSHEE NAVY DEPT., NATIONAL ARCHIVES (80-G-626130)

Aircraft of VF-12, Air Group One, based aboard the U.S.S. CORAL SEA (CVB-43). The color scheme and markings are appropriate for jet aircraft at the outbreak of the Korean War in mid 1950. The second squadron aboard a carrier was identified by the aircraft numbers being in the 200 block and also the white tip of the vertical tail.

4 SQUADRON TAIL MARKINGS

early 1930's

Naval Aviation had now reached the point where a visual aid was required for rapid recognition of various units oper--ating from the same base. A system of colored tail markings was gradually adopted for use by Observation, Scouting, Patrol, and Utility Squadrons.

The following pages illustrate, by squadron type, the colors and method of application used by various squadrons.

---------------- PATROL SQUADRONS ----------------

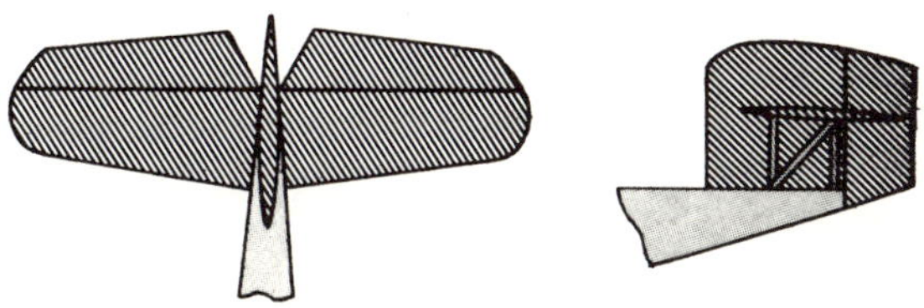

(Douglas PD-1 tail illustration)

Patrol Wing One, San Diego

VP-7	True Blue	1934 - 37
VP-9	Insignia Red	1934 - 37
VP-18 (VP-26)	Willow Green	1937 - 40 ..... transfered to PatWing 2 about the beginning of 1939.

Patrol Wing Two, Pearl Harbor

VP-1 (VP-21)	Insignia Red	1934 - 40
VP-4 (VP-22)	Lemon Yellow	1934 - 40
VP-6 (VP-23)	True Blue	1934 - 40
VP-8 (VP-24)	White	1934 - 40 ..... tail was left unpainted for a period of time during 1937 - 39.

Author's comments:

1. The number in brackets is for 1939 squadron re-designations.

solid tail color continued ...

Patrol Wing Four, Seattle

VP-20 (VP-44) True Blue 1937 - 40

Patrol Wing Five, Norfolk

VP-15 (VP-53) Insignia Red 1938 - 40

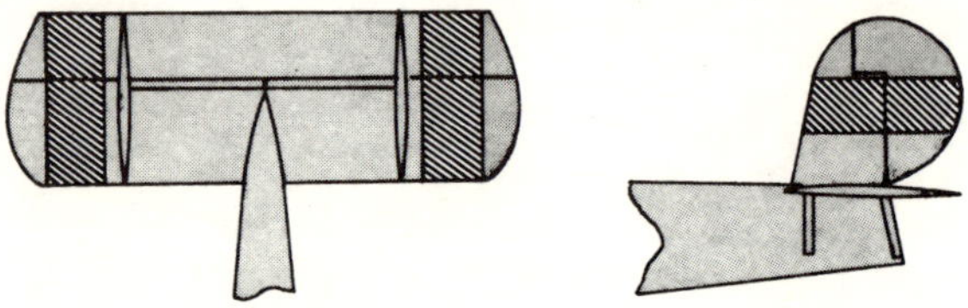

(Douglas P2D-1 tail illustration)

40" wide stripe

Patrol Wing Two, detached to Midway

VP-10 Black 1934 - 37

Patrol Wing Three, Coco Solo, Canal Zone

VP-2 (VP-31) White 1932 - 40 ..... tail color changed to willow green in 1937.

VP-3 (VP-32) Insignia Red 1932 - 40

VP-5 (VP-33) Black 1932 - 40

Patrol Wing Four, Seattle

VP-16 True Blue ____ - 37

VP-17 Insignia Red ____ - 37

Patrol Wing Five, Norfolk

VP-14 (VP-52) Lemon Yellow 1937 - 40

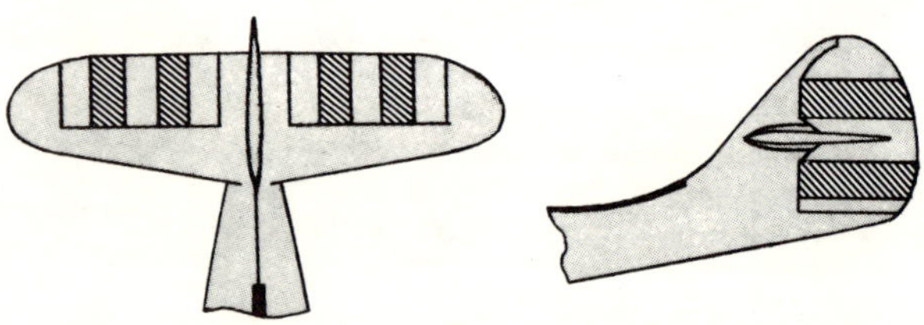

(Consolidated PBY-3 tail illustration)

24" wide stripes

Patrol Wing One, San Diego

VP-7 (VP-11) True Blue 1937 - 40

VP-9 (VP-12) Insignia Red 1937 - 40

VP-11 (VP-54) Willow Green 1937 - 40 ..... transfered to PatWing 5 about the end of 1938.

VP-12 (VP-51) Lemon Yellow 1937 - 40 ..... transfered to PatWing 5 about the end of 1938.

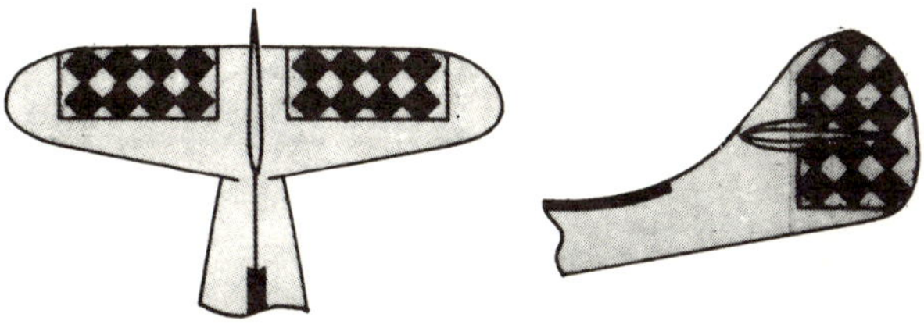

(Consolidated PBY-2 tail illustration)

30" squares reduced to 20" squares about 1939

Patrol Wing Two, detached to Midway

VP-10 (VP-25) Black 1937 - 40

Patrol Wing Four, Seattle and Alaska

VP-16 (VP-41) True Blue 1937 - 40

VP-17 (VP-42) Insignia Red 1937 - 40

VP-19 (VP-43) Lemon Yellow 1937 - 40

VP-21 (VP-14) Willow Green 1937 - 40 ..... transfered to PatWing 1 about the beginning of 1939.

<u>October 1940 -- February 1941</u>

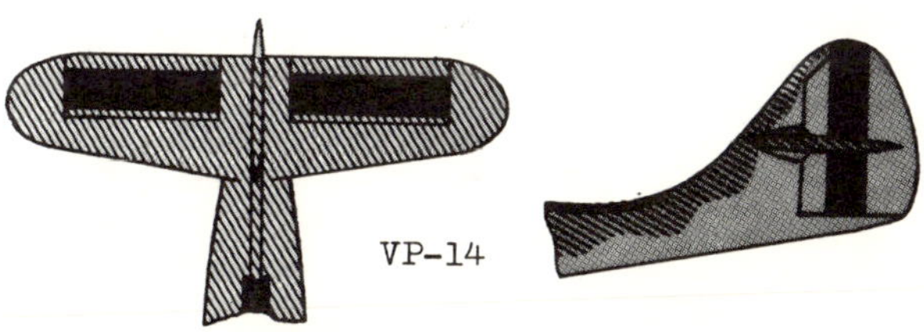

(Consolidated PBY-2 tail illustration)

Patrol Wing One	VP-11	Insignia Red
San Diego	VP-12	White
	VP-13	True Blue (late 1939 only)
	VP-14	Black

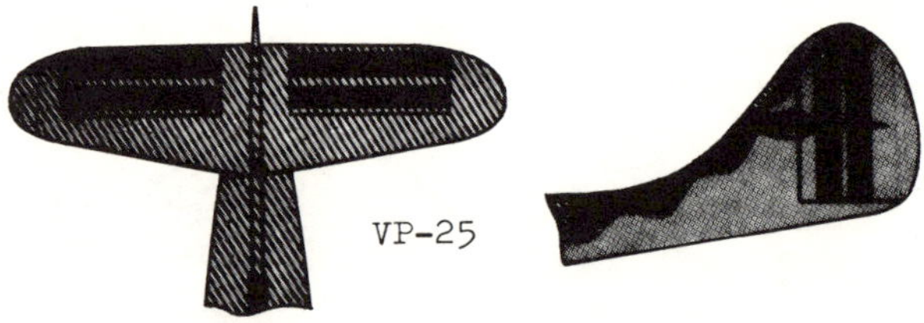

(Consolidated PBY-2 tail illustration)

Patrol Wing Two	VP-21	Red/White/Blue (detached to Cavit, Philippine Is.)
Pearl Harbor		
	VP-22	Insignia Red
	VP-23	White
	VP-24	True Blue
	VP-25	Black (detached to Midway)
	VP-26	Willow Green
	VP-27	Lemon Yellow
Patrol Wing Seven	VP-71	Insignia Red
	VP-72	White

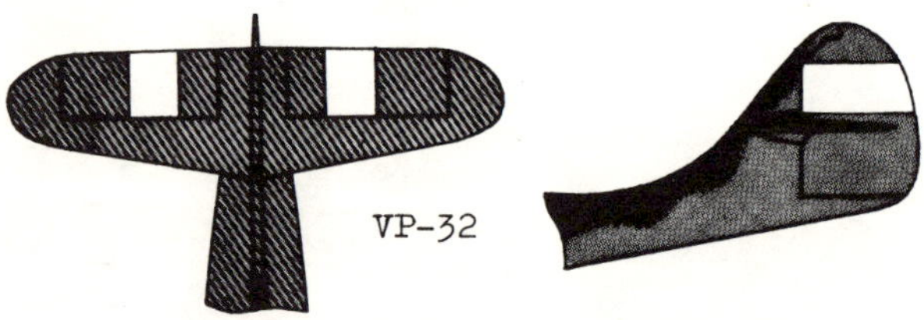

(Consolidated PBY-2 tail illustration)

Patrol Wing Three	VP-31	Insignia Red
Coco Solo, Canal Zone	VP-32	White
	VP-33	True Blue
	VP-34	Black

Author's comment:

Consolidated PBY Catalina aircraft had the colored tail markings applied to the moveable surfaces only.

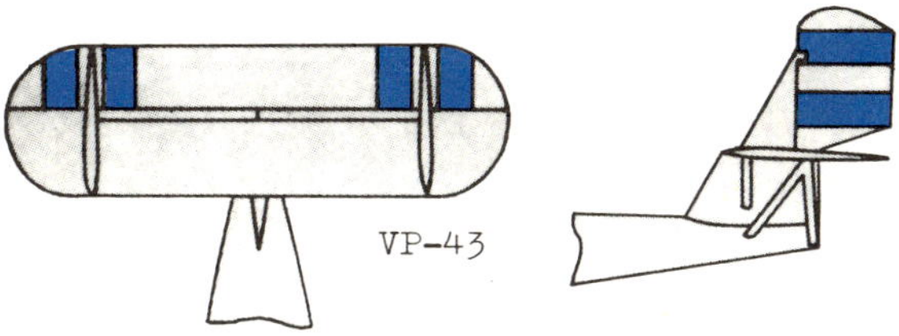

(Consolidated P2Y-3 tail illustration)

Patrol Wing Four	VP-41	Insignia Red
Seattle and Alaska	VP-42	White
	VP-43	True Blue
	VP-44	Black
	VP-45	Lemon Yellow (operated from Sitka, Alaska late 1939 only)

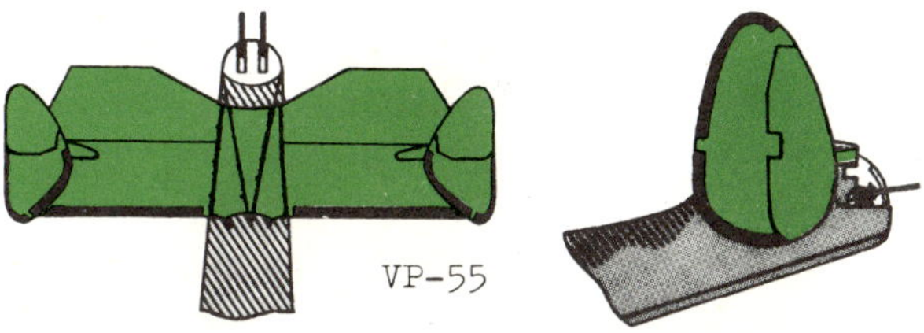

(Martin PBM-1 tail illustration)

Patrol Wing Five	VP-51	Insignia Red
Norfolk	VP-52	White
	VP-53	True Blue
	VP-54	Black
	VP-55	Willow Green
	VP-56	Lemon Yellow

Author's comment:

In late 1939 VP-52 turned their obsolete PB2Y's over to Training Command and new PBY's were transfered in from VP-14 at Seattle. The tail markings on these new aircraft were not changed to the squadron's assigned solid white tail.

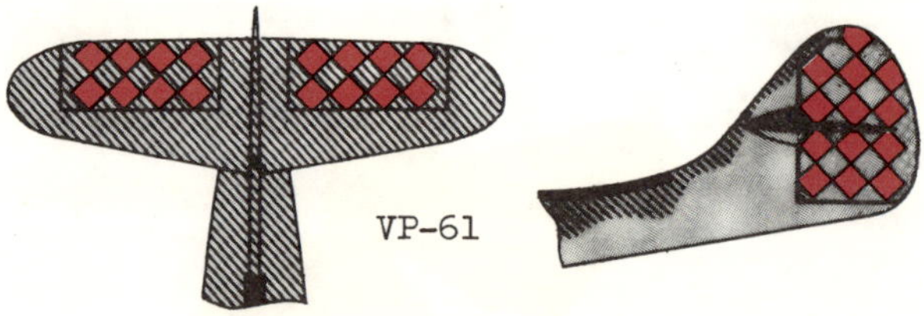

(Consolidated PBY-3 tail illustration)

Patrol Wing Six	VP-61	Insignia Red
	VP-62	White

BATTLESHIP OBSERVATION SQUADRONS

Battleship Division One	VO-1	(Insignia Red) 1933 - February 1941
Battleship Division Two	VO-2	(White) 1935 - February 1941
Battleship Division Three ...	VO-3	(True Blue) 1933 - February 1941
Battleship Division Four	VO-4	(Black) 1933 - February 1941
Battleship Division Five	VO-5	(Lemon Yellow) 1940 - February 1941

UTILITY SQUADRONS

VJ-1 (Willow Green) 1933 - February 1941
VJ-2 (Lemon Yellow) 1933 - February 1941
VJ-3 (Aluminum) 1940 - February 1941

CRUISER SCOUTING SQUADRONS

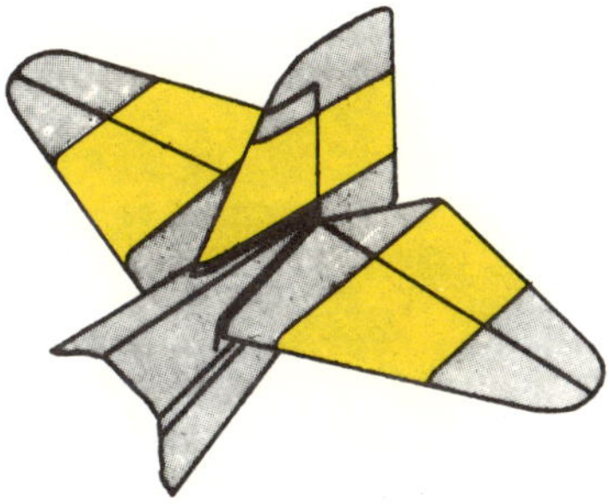

Cruiser Division Two ..... VS-5 (True Blue) 1933 - 1940
(re-numbered VCS-2 in 1937)
Operated two plane sections.

Cruiser Division Three ... VS-6 (Insignia Red) 1933 - 1940
(re-numbered VCS-3 in 1937)
Operated two plane sections.

Cruiser Division Four .... VS-9 (White) 1933 - 1940
(re-numbered VCS-4 in 1937)
Operated four plane sections.

Cruiser Division Five .... VS-10 (Lemon Yellow) 1933 - 1940
(re-numbered VCS-5 in 1937)
Operated four plane sections.

Cruiser Division Six ..... VS-11 (Willow Green) 1934 - 1937
(Black) 1937 - 1940
(re-numbered VCS-6 in 1937)
Operated four plane sections.

Cruiser Division Seven ... VS-12 (Black) 1935 - 1937
(Willow Green) 1937 - 1940
(re-numbered VCS-7 in 1937)
Operated four plane sections.

October 1940 - February 1941

Cruiser Division Four .... VCS-4 (True Blue) 4 plane sections
Cruiser Division Five .... VCS-5 (Lemon Yellow) 4 plane sections
Cruiser Division Six ..... VCS-6 (Black) 4 plane sections
Cruiser Division Seven ... VCS-7 (Willow Green) 4 plane sections

- -

October 1940 - February 1941

Cruiser Division Two ..... VCS-2 (True Blue) 2 plane sections

Cruiser Division Three ... VCS-3 (Insignia Red) 2 plane sections
Cruiser Division Eight ... VCS-8 (Black) 4 plane sections
Cruiser Division Nine .... VCS-9 (Willow Green) 4 plane sections

------------------ SHIP UNITS -------------------

October 1940 - February 1941

Destroyer Flotilla One ... U.S.S. RALEIGH (Insignia Blue)
Operated two aircraft.
Destroyer Flotilla Two ... U.S.S. DETROIT (Insignia Blue)
Operated two aircraft.
Submarine Force .... U.S.S. RICHMOND (Insignia Red)
Operated two aircraft.

Author's comments and additional marking information on this section:

a) The colored markings used by various squadrons were applied above and below the horizontal tail and/or elev--ators, and to both sides of the vertical tail and/or rudder.

b) The ship's name on aircraft from Battleship Observation

Squadrons and Cruiser Scouting Squadrons was carried in various locations below the cockpit. The location was finally standardized as indicated in the drawing below:

Location of ship's name as applied to aircraft from Cruiser Scouting Squadrons during 1933.

Location of ship's name as applied to aircraft from Cruiser Scouting Squadrons from 1934 to 1941.

Location of ship's name as applied to air--craft from Observation Squadrons from 1935 to 1941. Some aircraft carried the name a little lower and to the rear.

c) The aircraft tail illustrations in the Patrol Squadron portion indicate the aircraft in use by the illustrated squadron, in most cases, during the year the tail marking was adopted. The illustrations are not indicative of the aircraft in use by all squadrons carrying the same marking.

1947

Identification letters, 36 inches high, were applied to aircraft from Battleship Observation and Cruiser Scouting Squadrons. These letters were applied in white to both sides of the vertical tail, above the right wing, and below the left wing. The first letter, 'B' for battleship and 'C' for cruiser, indicated the squadron class and the second letter, assigned alphebetically, indicated the squadron within each class. Examples of these letters are shown below:

first battleship based squadron 'BA' (included U.S.S. IOWA)
sixth cruiser based squadron 'CF' (included U.S.S. TOPEKA)

Identification letters, 36 inches high, were also applied to aircraft of Patrol Squadrons. These letters were carried in white on both sides of the vertical tail, above the right wing, and below the left wing. The letters applied are shown below:

VPML-1 (VP-1).......AC (CD)
VPML-2 (VP-2).......SB
VPML-3 (VP-3).......CB
VPML-4 (VP-4).......SC
VPML-5 (VP-5).......MC
VPML-6 (VP-6).......BE
VPML-7 (VP-7).......HE
VPML-8 (VP-8).......HD
VPHL-10 (VP-20).....DD
VPHL-11 (VP-21).....HC
VPHL-2 (VP-22)......AE (CE)
VPHL-3 (VP-23)......EH
VPHL-4 (VP-24)......HA
VPHL-13 (VP-25).....BB
VPHL-6 (VP-26)......HB
VPHL-7 (VP-27)......DC
VPHL-8 (VP-28)......AF (CF)
VPHL-12 (VP-29).......DE
VPAM-2 (VP-32)........DB
VPAM-3 (VP-33)........EB
VPAM-4 (VP-34)........EC
VPMS-10 (VP-40).......CA
VPMS-11 (VP-41).......AA (MA)
VPMS-12 (VP-42).......SA
VPMS-3 (VP-43)........BC
VPMS-4 (VP-44)........CC
VPMS-5 (VP-45)........EE
VPMS-6 (VP-46)........AD (BD)
VPMS-7 (VP-47)........BA
VPMS-8 (VP-48)........MB
VPMS-9 (VP-49)........EA
VPW-1 (VP-51).........EW
VPP-1 (VP-61).........SD
VPP-2 (VP-62).........EF

( ) indicate changes as issued on 4 August 1948.

5 CARRIER SQUADRON COLORS

Late 1920's

The use of solid tail colors by Fighting and Bombing Squadrons was adopted in 1926. A partial list of tail colors, squadrons, and periods of service is shown below:

VF-1	Insignia Red	1926	Langley
VF-1B	Insignia Red	1927	Langley
" "	Insignia Red	1928 - March 1935	Saratoga
" "	Insignia Red	Apr '35 - Jun '36	Langley
VF-2	Insignia Red	1926	Langley
VF-2B	Lemon Yellow	1927 - 1930	Langley
" "	Lemon Yellow	1931, 1933 - 1937	Lexington
VF-3B	Willow Green	1928 - 1930	Lexington
" "	Willow Green	1932 - Jun 1934	Langley
" "	Willow Green	Jun '34 - Jun '36	Ranger
" "	White	Jun '36 - Jun '37	Saratoga
VF-3S	Insignia Red	1931	Langley
VF-5B	True Blue	1930 - 1931	Lexington
" "	True Blue	1933 - 1935	Lexington
" "	True Blue	1936 - 1937	Ranger

VF-5S	True Blue	1932	Lexington
VF-6B	White	1927	Langley
" "	White	1930 - 1937	Saratoga
VB-1B	Lemon Yellow	Jul '28 - 1929	Lexington
" "	True Blue	1929	Lexington
" "	White	Jun '36 - Jun '37	Saratoga
VB-2B	White	1928 - 1929	Saratoga
" "	Insignia Red	1935 - 1937	Saratoga
VB-3B	Willow Green	1936 - 1937	Lexington
VB-5B	Willow Green	Nov '34 - Jun '35	Ranger
" "	Willow Green	Jun '35 - 1937	Lexington

July 1st 1937

The use of solid tail colors by individual carrier squad--rons ceased and the colors were assigned to the carriers.

Author's comments:

a) Squadrons, especially Fighting and Bombing Squadrons, were re-designated quite frequently. An example is Fighting Four (VF-4); which began life as Fighting Five (VF-5), changed to Bombing One (VB-1), changed back to Fighting Five, and finally became Fighting Four. Fighting Four then went on to become Fighting Forty-One (VF-41), the first of two fighting squadrons based aboard the U.S.S. RANGER. The second squadron on the RANGER being Fighting Forty-Two (VF-42).

Some squadrons retained the same tail colors throughout these changes and others did not; for this reason I have not attempted to trace squadrons and tail colors as this would take a small book in itself.

Color coding was authorized for use by the six squadrons in a Carrier Air Group on the 1st of May. The colors were applied to the spinner and to both sides of the vertical tail and rudder.[1]

By this time a new aircraft numbering system had been adopted for use. This system used three numbers..... the first number, 1 through 6, indicated the squadron within the Air Group and the last two numbers indicated the aircraft within the squadron.

The squadron colors and numbering system are shown below:

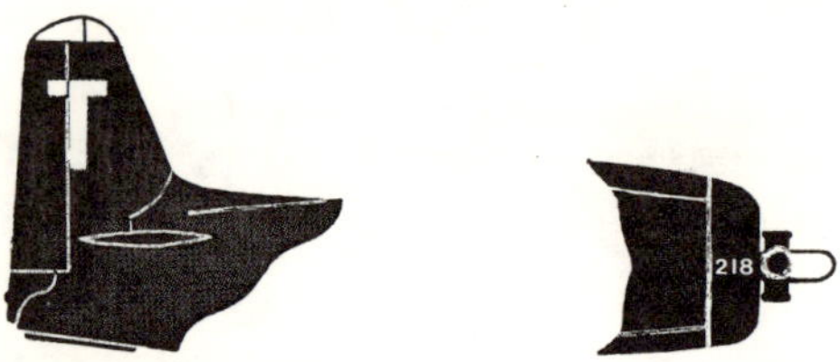

aircraft from the second squadron (VF-12)
assigned to Carrier Air Group One (CVG-1)

1st squadron .... 101 up .... Insignia Red
2nd squadron .... 201 up .... Insignia White
3rd squadron .... 301 up .... Light Blue
4th squadron .... 401 up .... Light Yellow
5th squadron .... 501 up .... Light Green
6th squadron .... 601 up .... Black outlined in white

1. It appears as though jet aircraft did not normally carry the squadron color on the nose of the aircraft.

1950

During the Korean War, individual squadron colors and art decoration began to appear. Squadron Commanders were permitted to use red and white for these art designs. Examples of the art decorations, usually carried on the nose of the aircraft, are shown below.

Fighting Ninety-One (VF-91), Air Group 9 aboard the U.S.S. PHILIPPINE SEA, Korea, 1953.

(The assigned Air Group Identification Letter 'N' was carried on the tail and wings as normal.)

Fighting One Ninety-One (VF-191), Air Group 19 aboard the U.S.S. PRINCETON, Korea, 1951.

(The assigned Air Group Identification Letter 'B' was carried on the wings and tail as normal.)

Fighting One Twelve (VF-112), Air Group 11 aboard the U.S.S. PHILIPPINE SEA, Korea, 1951.

(The assigned Air Group Identification Letter 'V' was carried as normal on the wings and tail.)

6 CARRIER IDENTIFICATION MARKINGS

1937

Prior to July, carriers had used tail colors that were devised by Carrier Commanders -- e.g. U.S.S. LANGLEY used red and U.S.S. LEXINGTON used yellow. Carrier based VF and VB squadrons used their own tail colors, and as the number of carriers and squadrons increased, so did the confusion. The Navy now decided to standardize the colors and eliminate this perplexity. On July 1st the use of tail colors by VF and VB squadrons ceased and the colors were assigned to the carriers instead; now all squadrons would carry the tail color assigned to the carrier they were based aboard. The colors assigned and method of application are shown below:

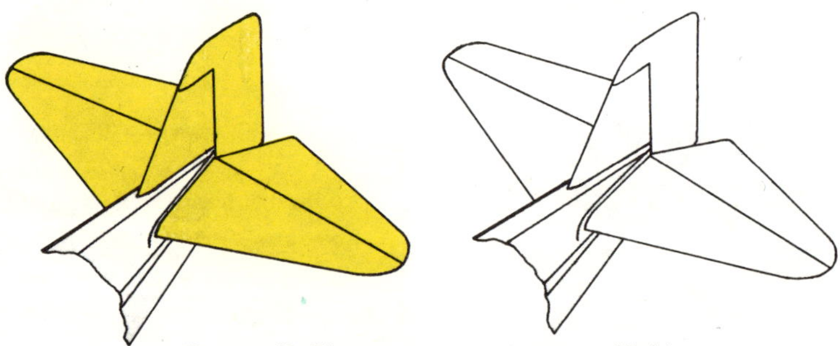

Lemon Yellow
U.S.S. LEXINGTON (CV-2)

White
U.S.S. SARATOGA (CV-3)

Author's comments:

a) The U.S.S. LANGLEY (CV-1) had used red as her tail color prior to July, 1937; when she was assigned as a seaplane tender, the U.S.S. YORKTOWN (CV-5) took the LANGLEY's red tail color. The newly commissioned U.S.S. WASP (CV-7) then inherited the black tail color which was originally assigned to the YORKTOWN.

Willow Green
U.S.S. RANGER (CV-4)

Insignia Red
U.S.S. YORKTOWN (CV-5)

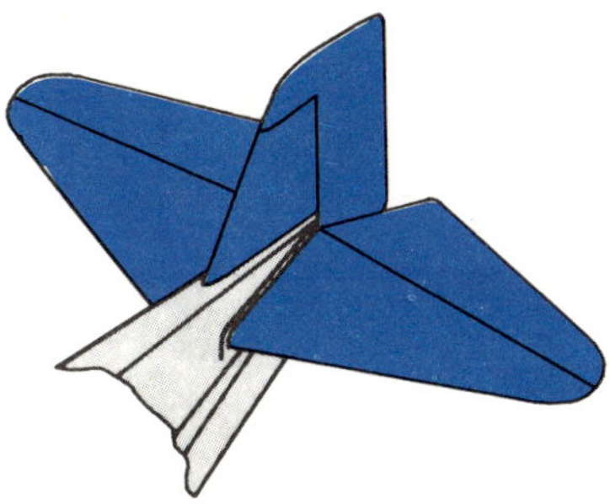

True Blue
U.S.S. ENTERPRISE (CV-6)

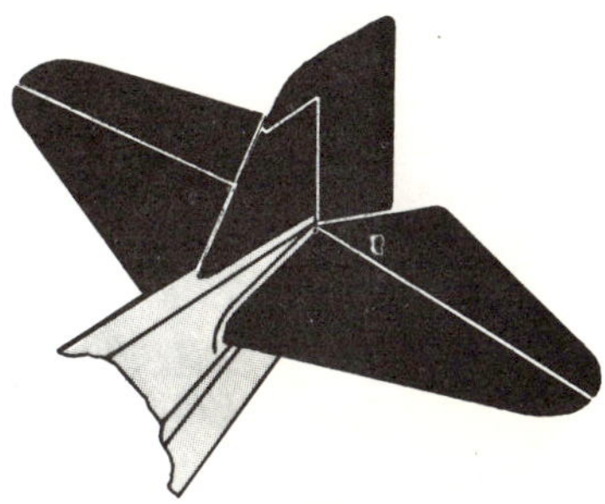

Black
U.S.S. WASP (CV-7)

1941

The use of colored tail markings was discontinued on the 26th of February.

1943

The increase in the number of aircraft carriers created the need for additional identification markings on carrier based aircraft. By the end of the year an unofficial system of markings was in use by many of the carriers. These markings, applied in white, are shown on the following pages.

1

2

3

1. U.S.S. ENTERPRISE (CV-6), Air Group 20 (CVG-20)
2. U.S.S. ESSEX (CV-9), Air Group 15 (CVG-15)
3. U.S.S. YORKTOWN (CV-10), Air Group 1 (CVG-1) *

*(A green propeller spinner was also a YORKTOWN marking)
*(VF-1 carried the letter 'K' during May - July 1944)

4
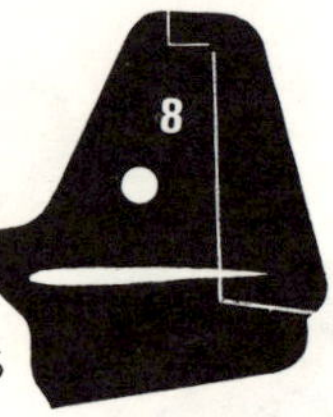

5

6

4. U.S.S. INTREPID (CV-11), Air Group 18 (CVG-18) *
5. U.S.S. HORNET (CV-12), Air Group **11** (CVG-11)
6. U.S.S. TICONDEROGA (CV-14), Air Group 80 (CVG-80)

*(A small white horizontal bar, under the aircraft number, was carried until the 'cross' was adopted in mid-1944)

7
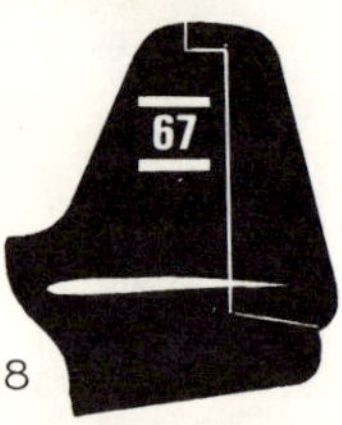

8

9

7. U.S.S. LEXINGTON (CV-16), Air Group 19 (CVG-19) *
8. U.S.S. BUNKER HILL (CV-17), Air Group 8 (CVG-8)
9. U.S.S. WASP (CV-18), Air Group 14 (CVG-14)

*(VF-19 did not carry the 'hollow triangle' marking.)

10

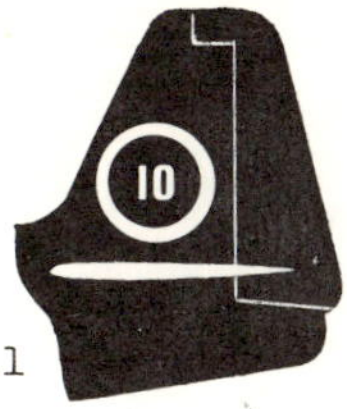

11

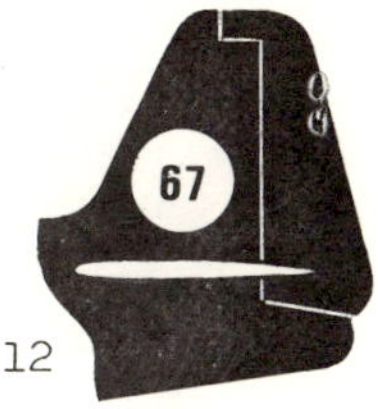

12

10. U.S.S. HANCOCK (CV-19), Air Group 7 (CVG-7)
11. U.S.S. INDEPENDENCE (CVL-22), Air Group (Night) 41 (CVG(N)-41)
12. U.S.S. BELLEAU WOOD (CVL-24), Air Group 21 (CVG-21)

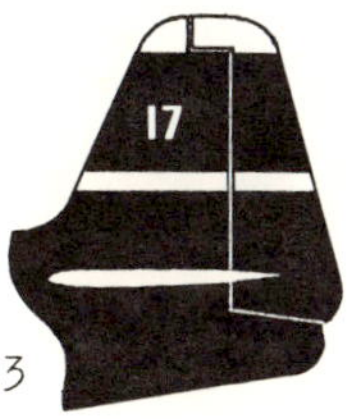

13

14

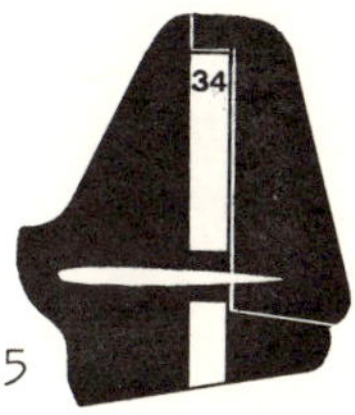

15

13. U.S.S. COWPENS (CVL-25), Air Group 22 (CVG-22)
14. U.S.S. LANGLEY (CVL-27), Air Group 44 (CVG-44)
15. U.S.S. CABOT (CVL-28), Air Group 29 (CVG-29)

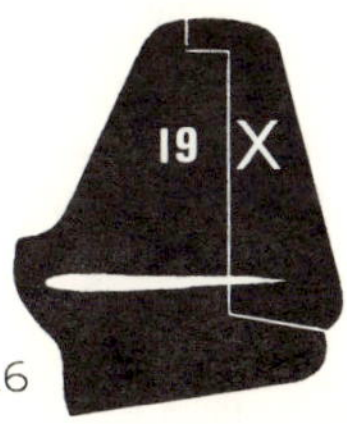

16

17

16. U.S.S. SAN JACINTO (CVL-30), Air Group 51 (CVG-51)
17. U.S.S. SARATOGA (CV-3), Air Group 12 (CVG-12) *
 *(white stripes are on top of horizontal tail surfaces only.)

18. **U.S.S. MONTEREY** (CVL-26), Air Group 28 (CVG-28) *

*(white wing marking was carried above and below both wing tips)

19. **U.S.S. FRANKLIN** (CV-13), Air Group 13 (CVG-13)
    no markings
20. **U.S.S. PRINCETON** (CVL-23), Air Group 27 (CVG-27)
    no markings (See sharkmouth marking in Section 8)
21. **U.S.S. BATAAN** (CVL-29), Air Group 50 (CVG-50)
    no markings
22. **U.S.S. RANGER** (CV-4)
    used to train carrier pilots - markings were not required

1944

An order was issued on the 7th of October which assigned specific markings to all CV and CVL class carriers. This order gave official sanction to the unofficial practice which had been in use since early 1943. Some of these markings were never applied, and others were changed in actual use because of a carrier's role. These markings and their method of appli--cation are shown on the following pages. It appears as though most carriers did not adopt these new markings until January 1945.

Author's comments:

a) The aircraft numbers used in these drawings are examples only, used to show the location of the numbers as applied by various carriers.

b) Air Group numbers are for the late 1944 period.

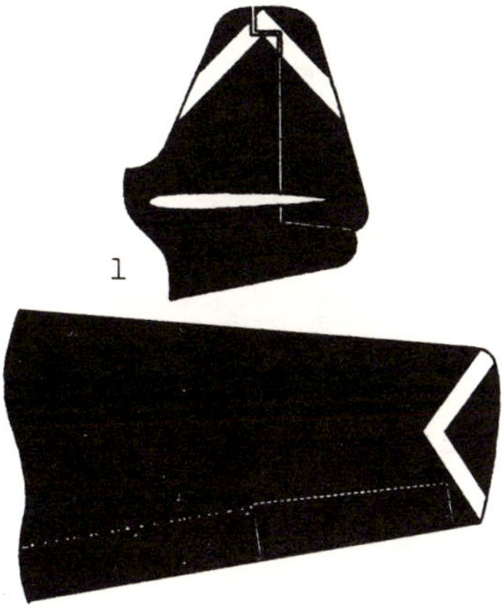

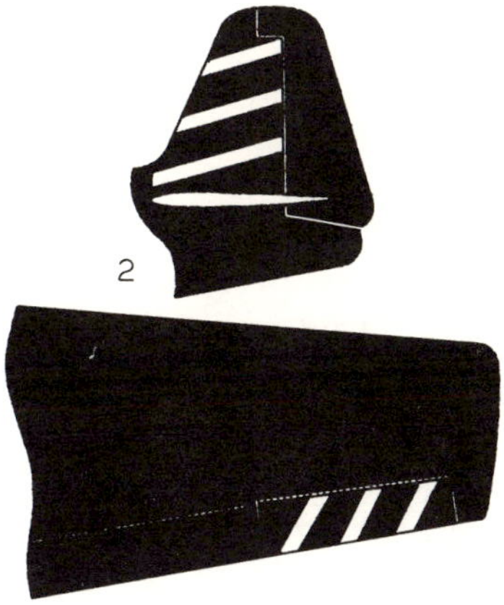

1. U.S.S. SARATOGA
   (CV-3)

Marking on upper right wing only.

2. U.S.S. ENTERPRISE
   (CV-6)

Original markings as assigned.

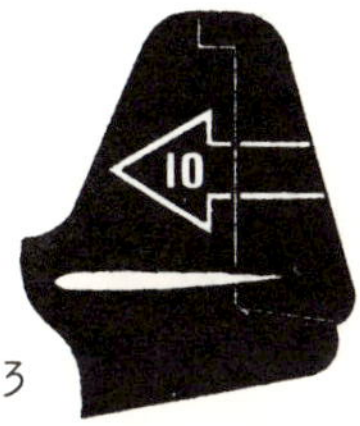

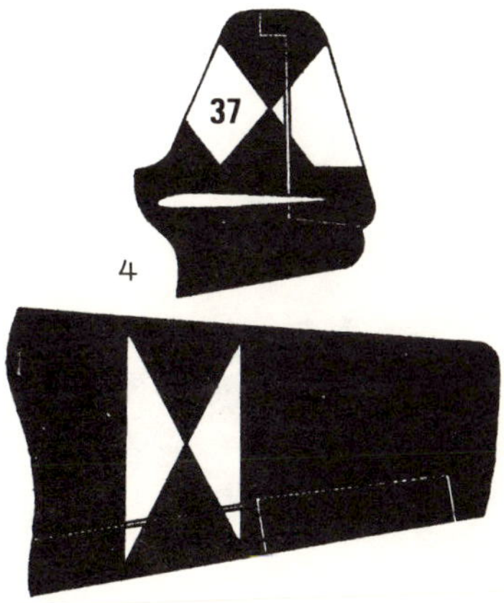

3. U.S.S. ENTERPRISE
   (CV(N)-6)

No wing markings.

Markings adopted for use because the 'Big E' was operating as a night carrier.

4. U.S.S. ESSEX
   (CV-9)

Marking on upper right wing only.

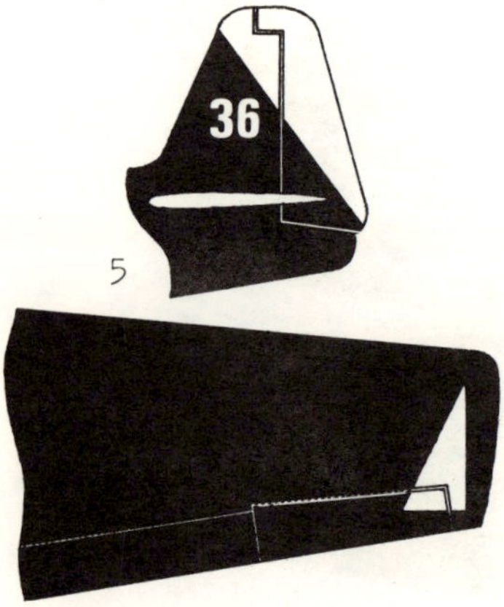

5

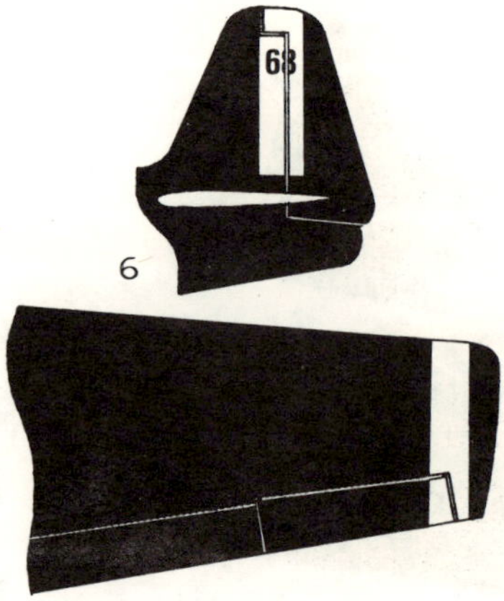

6

5. U.S.S. YORKTOWN
(CV-10)
Marking on upper right wing only. Green spinner.

6. U.S.S. INTREPID
(CV-11)
Marking on upper right and lower left wing.

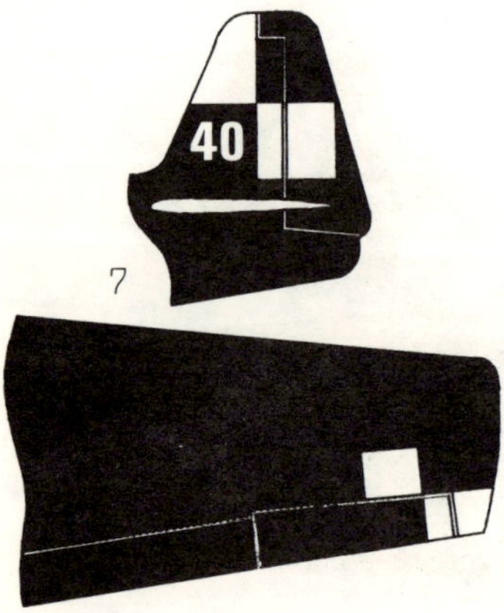

7

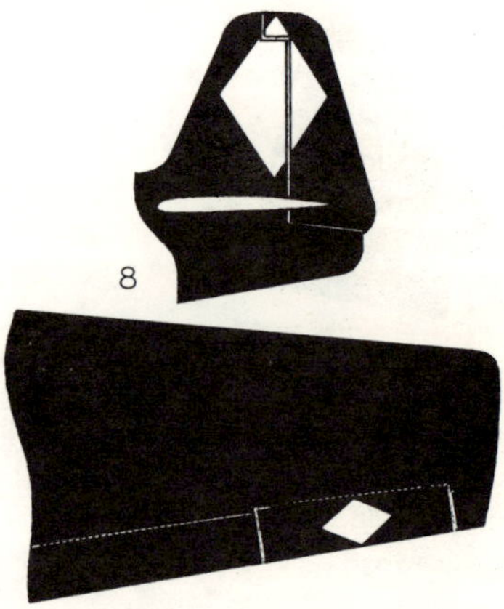

8

7. U.S.S. HORNET
(CV-12)
Marking on upper right and left wing. White spinner.

8. U.S.S. FRANKLIN
(CV-13)
Marking on upper right and lower left wing.

9

9. U.S.S. TICONDEROGA
(CV-14)

Marking on upper right and lower left wing.

10

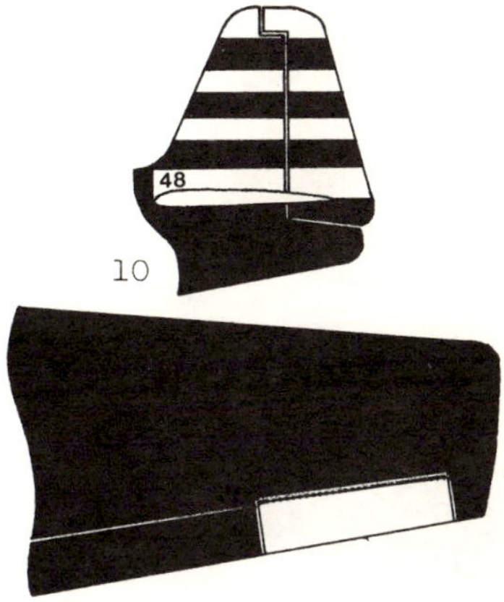

10. U.S.S. RANDOLPH
(CV-15)

Marking on upper and lower left and right wing.

11

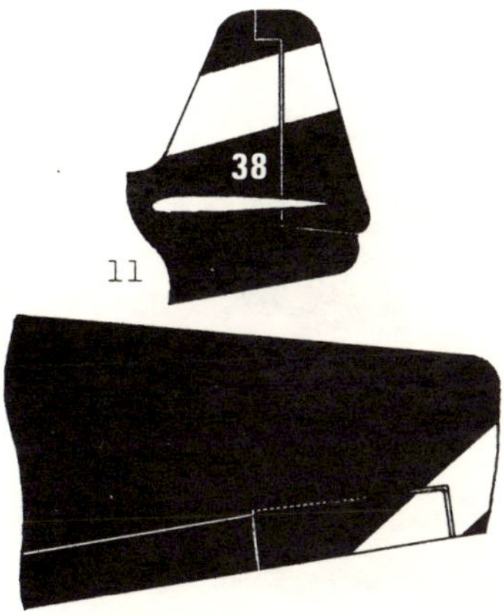

11. U.S.S. LEXINGTON
(CV-16)

Marking on upper right and left wing; moved to lower left in June, 1945.

12

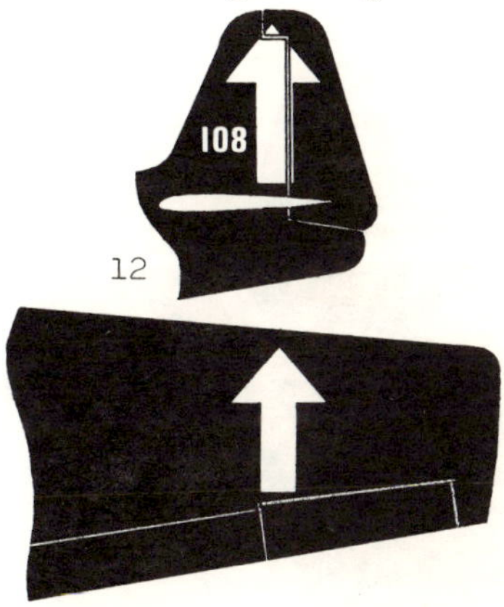

12. U.S.S. BUNKER HILL
(CV-17)

Marking on upper right and lower left wing.

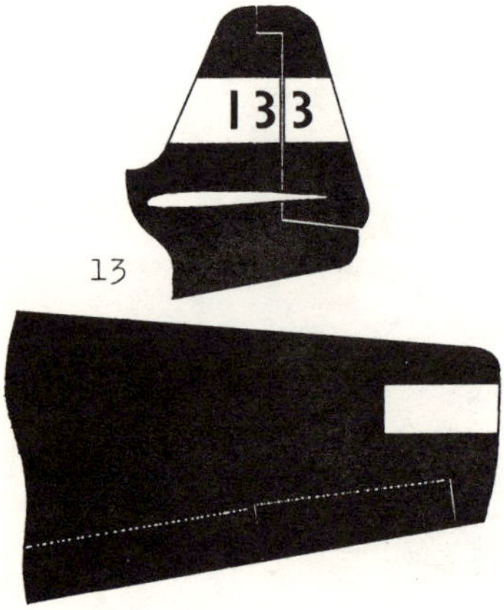

13. U.S.S. WASP
(CV-18)
Marking on upper right wing only. Early 1945 added under left wing.

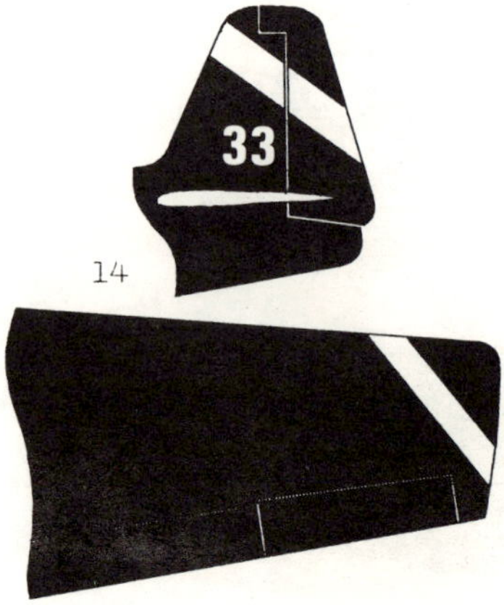

14. U.S.S. HANCOCK
(CV-19)
Marking on upper right wing only. White spinner.

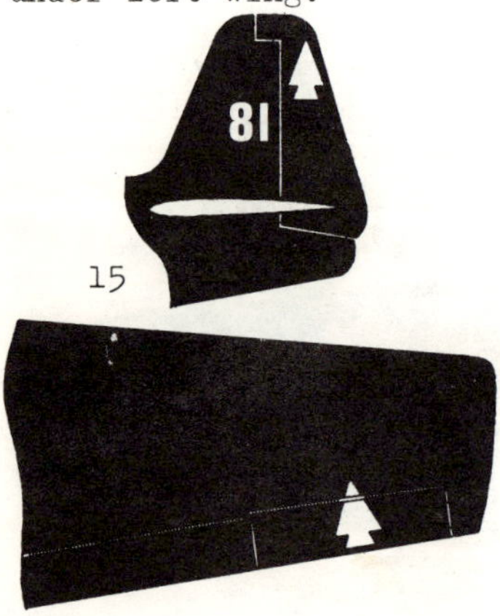

15. U.S.S. BENNINGTON
(CV-20)
Marking on upper and lower left and right wing.

16. U.S.S. BOXER
(CV-21)
Marking on upper right wing only.

17

18

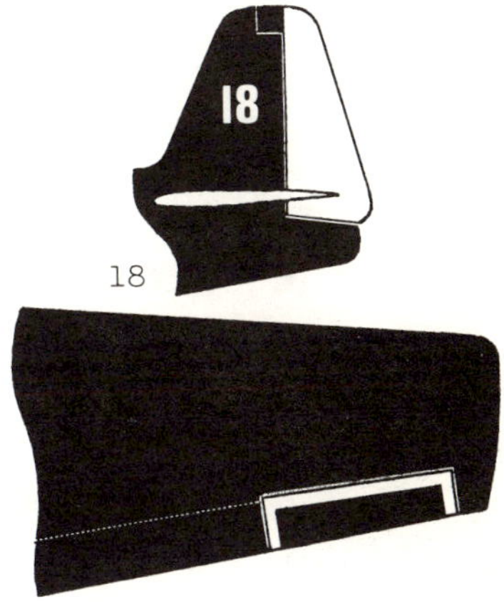

17. **U.S.S. INDEPENDENCE**
(CVL-22)
Marking on upper and lower right wing only.

18. **U.S.S. BELLEAU WOOD**
(CVL-24)
Marking on upper right and left wing.

19

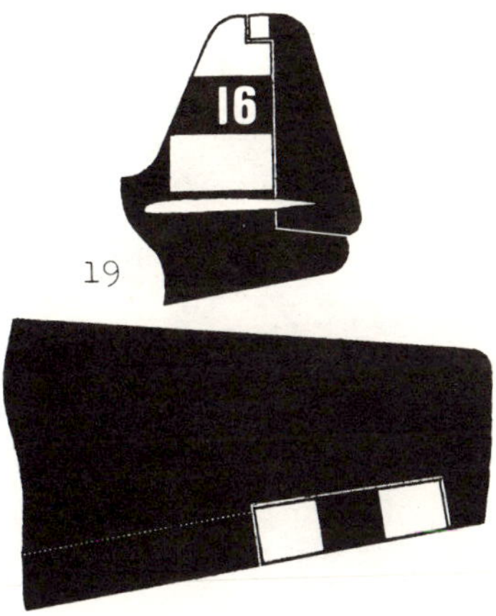

20

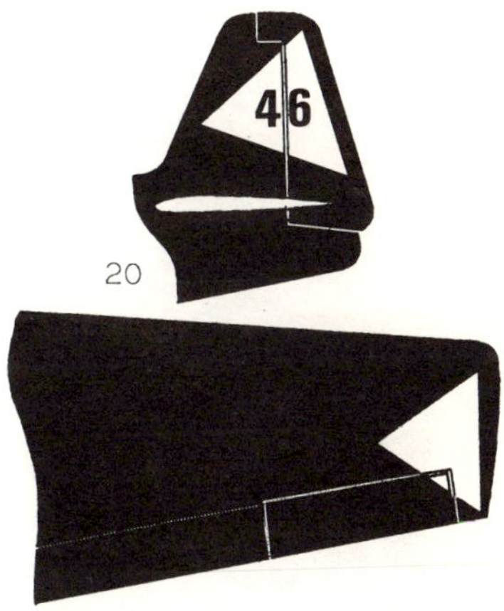

19. **U.S.S. COWPENS**
(CVL-25)
Marking on upper right wing only.

20. **U.S.S. MONTEREY**
(CVL-26)
Marking on upper right and lower left wing.

21

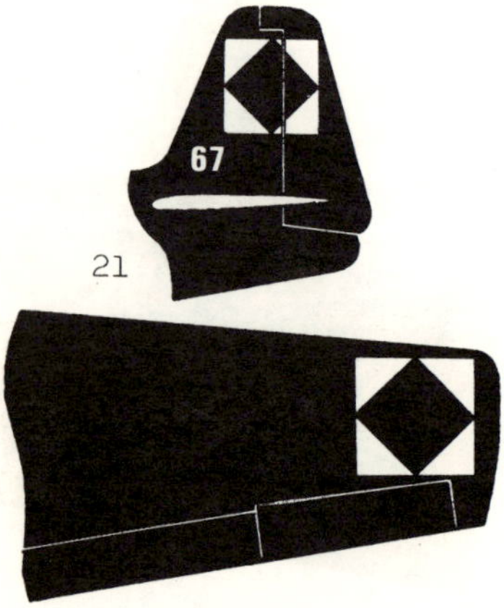

22

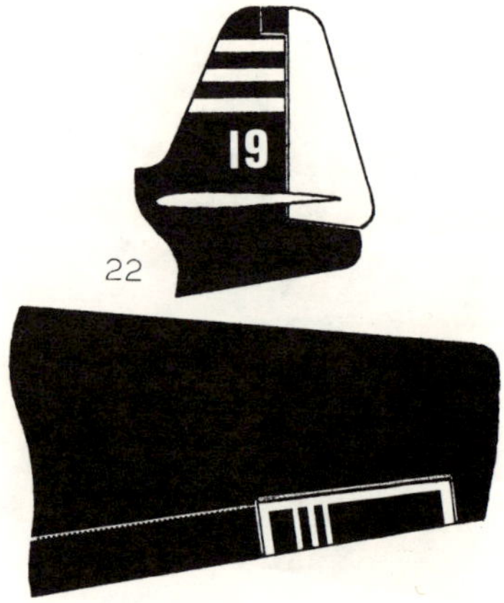

21. U.S.S. LANGLEY
(CVL-27)
Marking on upper right wing only.

22. U.S.S. CABOT
(CVL-28)
Marking on upper right and left wing.

23

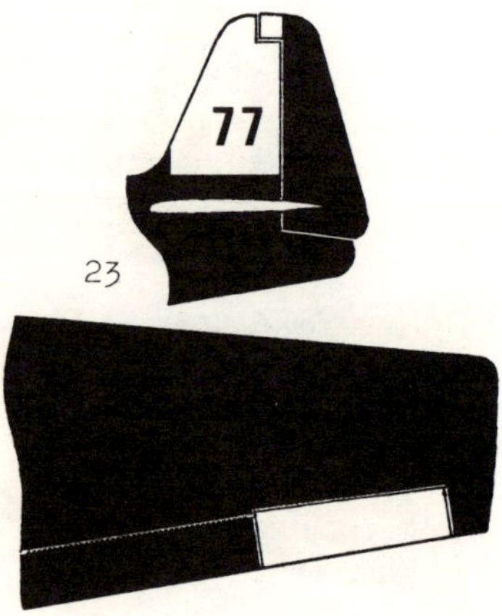

24

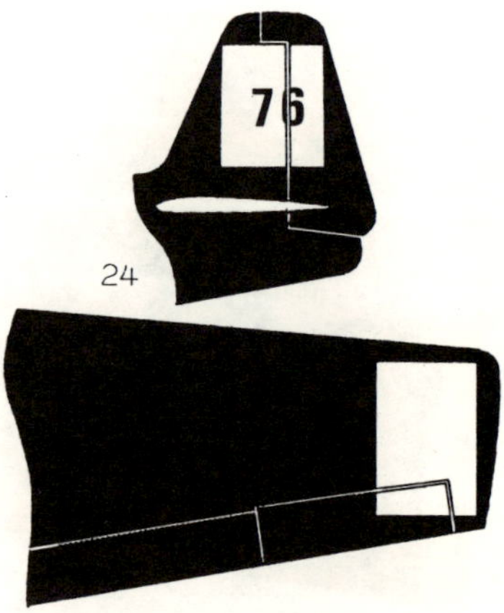

23. U.S.S. BATAAN
(CVL-29)
Marking on upper and lower left and right wing.

24. U.S.S. SAN JACINTO
(CVL-30)
Marking on upper right wing only.

25

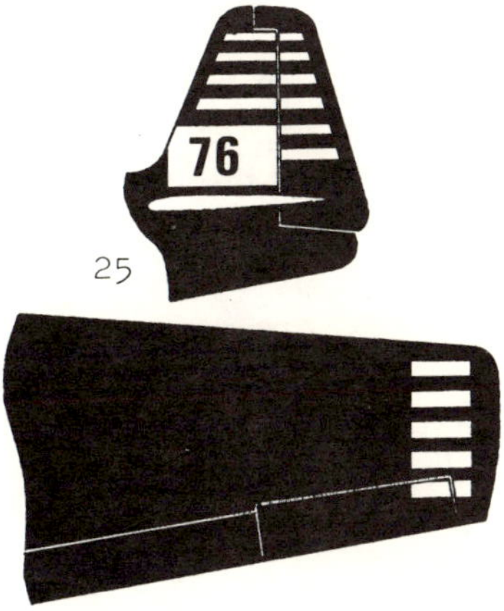

25. U.S.S. BON HOMME RICHARD
(CV-31)
Marking on upper right wing only.

26

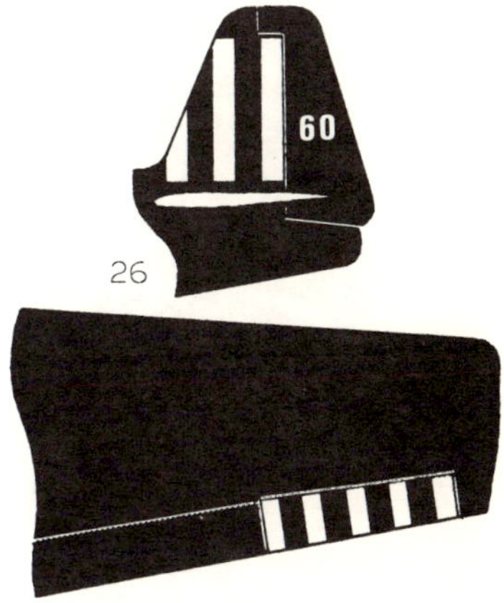

26. U.S.S. ANTIETAM
(CV-36)
Marking on upper right and left wing.

27

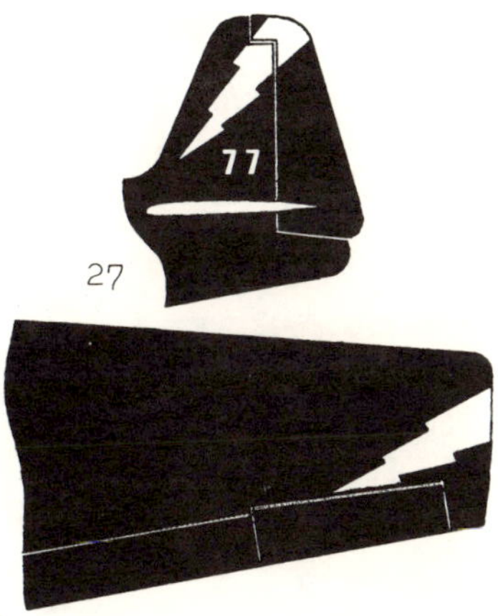

27. U.S.S. SHANGRI LA
(CV-38)
Marking on upper right and lower left wing.

28

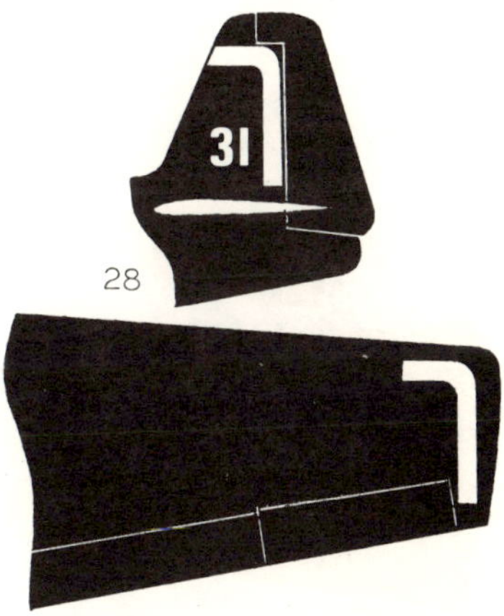

28. U.S.S. LAKE CHAMPLAIN
(CV-39)
Marking on upper right and left wing.

1945

A new system of double or single letters was adopted for use on 27 July. These letters were to be at least 24 inches high and were applied in white to both sides of the vertical tail, upper right, and lower left wing tips.

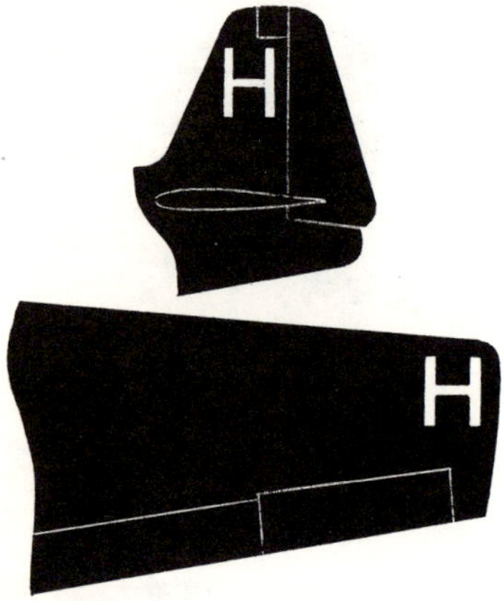

U.S.S. Lexington
(CV-16)

U.S.S. Saratoga (CV-3)........................ CC
U.S.S. Ranger (CV-4) .... PP
U.S.S. Enterprise (CV-6) .... M
U.S.S. Essex (CV-9) .... F
U.S.S. Yorktown (CV-10) .... RR
U.S.S. Intrepid (CV-11) .... E
U.S.S. Hornet (CV-12) .... S
U.S.S. Franklin (CV-13) LL
U.S.S. Ticonderoga (CV-14) V
U.S.S. Randolph (CV-15) L

Author's comments:

a) The change over to the new markings was gradual and some carriers retained the 'Geometric' symbols until the end of the war, while others used a combination of both markings.

U.S.S. Bunker Hill (CV-17) .... Y
U.S.S. Wasp (CV-18) X
U.S.S. Hancock (CV-19) .... U
U.S.S. Bennington (CV-20) .... TT
U.S.S. Boxer (CV-21) .... ZZ
U.S.S. Independence (CVL-22) .... D
U.S.S. Belleau Wood (CVL-24) .... P
U.S.S. Cowpens (CVL-25) .... A
U.S.S. Monterey (CVL-26) C
U.S.S. Langley (CVL-27) .... K
U.S.S. Cabot (CVL-28) .... R
U.S.S. Bataan (CVL-29) .... T
U.S.S. San Jacinto (CVL-30) .... B
U.S.S. Bon Homme Richard (CV-31) .... SS
U.S.S. Antietam (CV-36) .... W
U.S.S. Shangri La (CV-38) .... Z
U.S.S. Lake Champlain (CV-39) .... AA
U.S.S. Midway (CVB-41) .... YY
U.S.S. Franklin D. Roosevelt (CVB-42) .... FF
U.S.S. Coral Sea (CVB-43) .... EE

1946

November 7 — New Identification Letters were assigned to all carriers. This system of single or double letters was designed to indicate the name of the carrier. If the first letter of the carrier's name was not duplicated by another carrier, then the first letter only was used. If the first letter was duplicated by another carrier, then the first two letters were used. If the carrier's name was two words, then the first letter of each word was used. The letters G, J, O, Q, and Y were not used; if a carrier's name began with one of the letters, then the second or second and third letters were used.

December 12 — The assignment of ID letters to individual carriers was discontinued and the letters assigned to the Air Groups aboard the carriers instead. If the Air Group was transfered to another carrier it would retain the letters already assigned.

1947

January 2 — Identification letters, as assigned in December, were ordered displayed on both sides of the vertical tail and rudder, and on the upper right and lower left wing tips. The letters applied are shown below.

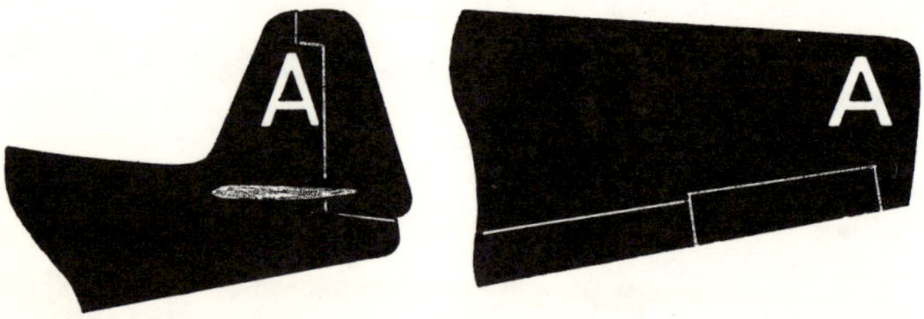

U.S.S. Antietam (CV-36) .... CVG-15 ... A

U.S.S. Boxer (CV-21) .... CVG-19 ... B
U.S.S. Coral Sea (CVB-43) .... CVBG-5 ... C ... (CVG-6)
U.S.S. Franklin D. Roosevelt (CVB-42) .. CVBG-3 ... F ... (CVG-4)
U.S.S. Kearsarge (CV-33) .... CVG-3 .... K
U.S.S. Leyte (CV-32) .... CVG-7 L
U.S.S. Midway (CVB-41) .... CVBG-1 ... M ... (CVG-2)
U.S.S. Princeton (CV-37) .... CVG-13 ... P
U.S.S. Philippine Sea (CV-47) .... CVG-9 .... PS .. (D)
U.S.S. Randolph (CV-15) .... CVG-17 ... R
U.S.S. Oriskany (CV-34) .... RI
U.S.S. Shangri La (CV-38) .... CVG-5 .... S

U.S.S. Saipan (CVL-48) .... CVLG-1 .. SA .. (VX-3) .. (XC)
U.S.S. Tarawa (CV-40) .... CVG-1 .... T
U.S.S. Valley Forge (CV-45) .... CVG-11 ... V

Author's comments:

a) The ( ) indicate changes which were issued in August 1948.

b) The above list was prepared from Aviation Circular Letter No. 69-48 which was issued on 4 August 1948, and indicated changes in Air Group numbers and letters. Also used were published articles and correspondence which included a few examples of the assigned letters.

There may be errors, and possibly omissions, in the above list, however, it is as complete and accurate as I can make it at this time.

7 ESCORT CARRIER MARKINGS

1944 - 1945

The increase in the number of aircraft carriers created the need for identification markings on CVE based aircraft. During this period many carriers began to apply markings that were devised by the Carrier Commanders. These markings were carried in white on both sides of the vertical tail. A few examples of these markings are shown below.[1]

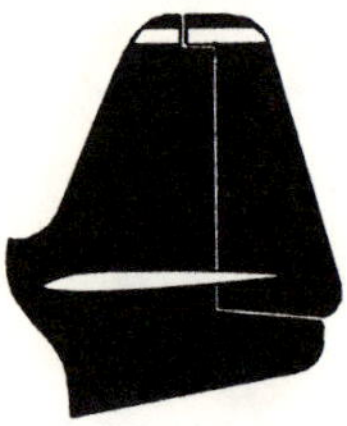

U.S.S. ALTAMAHA (CVE-18)

U.S.S. SANGAMON (CVE-26) *

*(Changed to small bar under aircraft number in early 1945.)

U.S.S. HOGGATT BAY (CVE-75)

U.S.S. RUDYERD BAY (CVE-81)
(applied using 3" masking tape)

1. There were undoubtedly more of these markings, however, at this time I have not located any. If any reader can shed some light on this section, I would be most grateful.

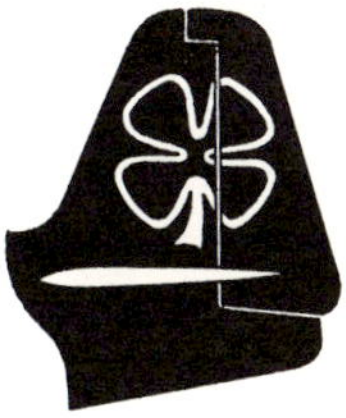

U.S.S. SHAMROCK BAY (CVE-84) U.S.S. MAKASSAR STRAIT (CVE-91)

(Written reference has been found indicating that Shamrock Bay carried a white shamrock with crossed Irish pipes underneath. The illustrated marking was taken from a photograph.)

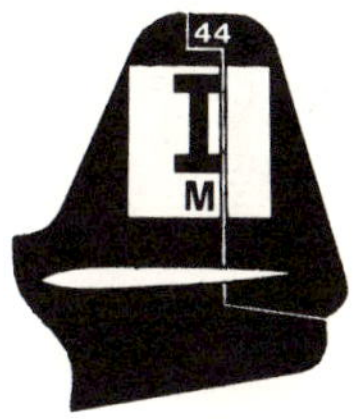

U.S.S. BLOCK ISLAND (CVE-106) U.S.S. BLOCK ISLAND (Mar '45)

(Block Island was the first of four escort carriers to carry only Marine Corps squadrons, embarking in March 1945 with VMTB-233 aboard. The other carriers and squadrons were as follows: VMTB-143 aboard the U.S.S. GILBERT ISLANDS (CVE-107), VMTB-234 aboard the U.S.S. VELLA GULF (CVE-111), and VMTB-132 aboard the U.S.S. CAPE GLOUCESTER (CVE-109).)

June 1945

A directive was issued on the 2nd of June which assigned new wing and tail markings to CVE based aircraft. This action was taken to standardize the markings in use and to facilitate identification of aircraft operating from escort carriers.

These markings were applied to both sides of the rear fuse-lage and vertical tail, and on the upper right wing surface.

The markings were applied in white, with a second color being used in some of the markings.[1]

The tail and wing tip markings indicated the Carrier Div--ision to which the carrier was attached, and the stripes des--ignated the individual carriers within the Carrier Division.

CARRIER DIVISION TWENTY-TWO

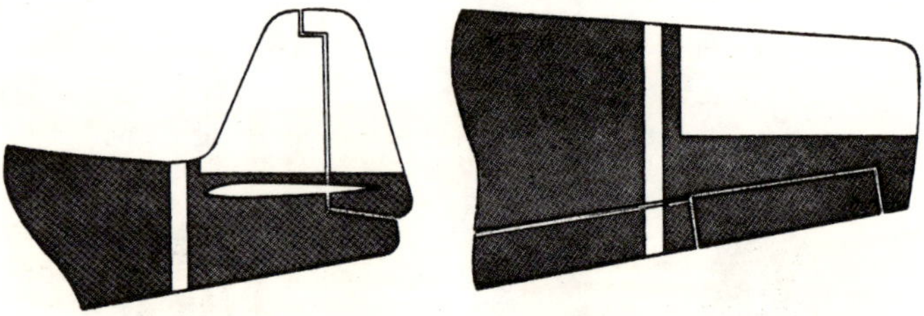

U.S.S. SANGAMON (CVE-26)

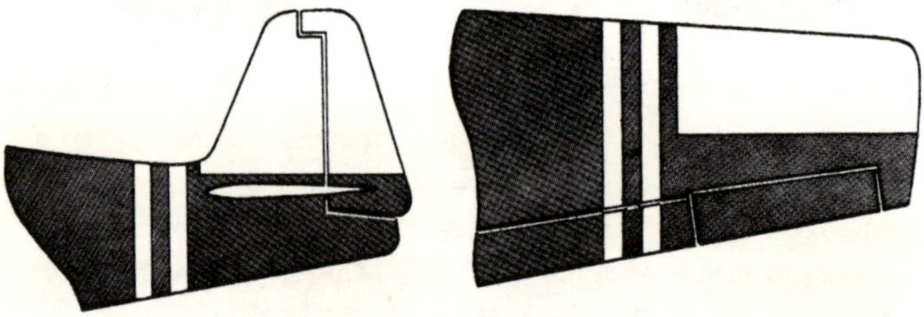

U.S.S. SUWANEE (CVE-27)

1. The second color used may have been light yellow or light green, as these colors had previously been called for in temporary markings. I assume yellow would have been used, as it contrasted with the sea blue over-all finish which was in use at that time.

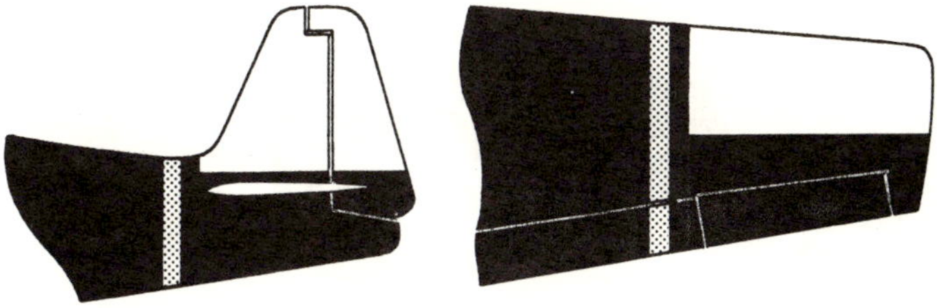

U.S.S. CHENANGO (CVE-28)

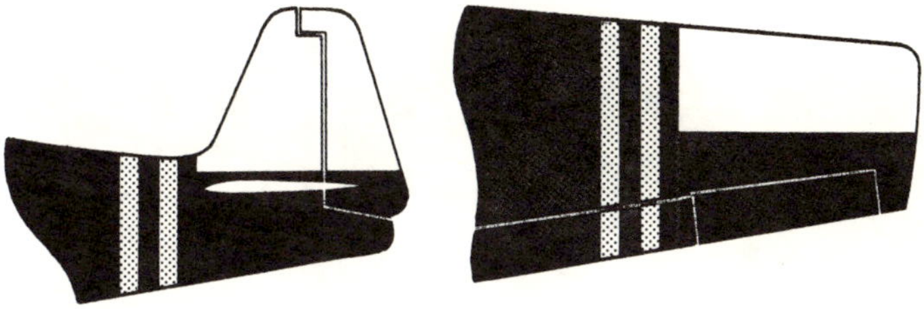

U.S.S. SANTEE (CVE-29)

CARRIER DIVISION TWENTY-THREE

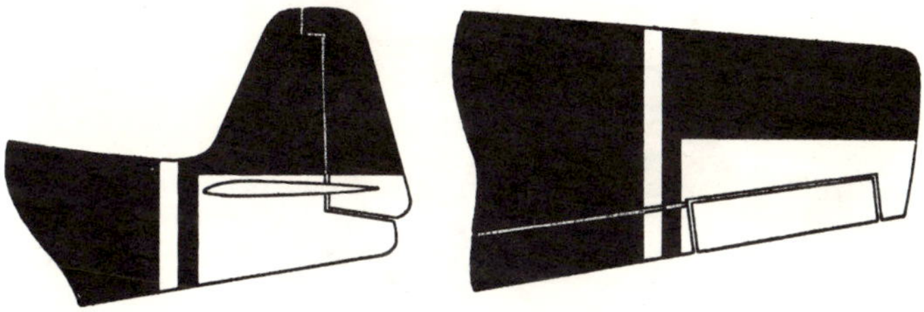

U.S.S. KITKUN BAY (CVE-71)

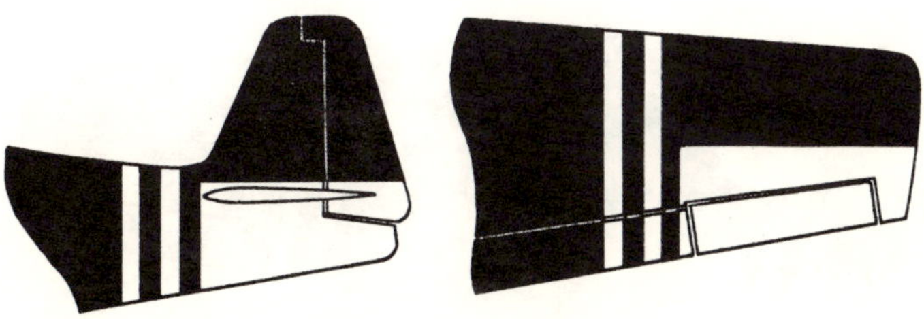

U.S.S. HOGGATT BAY (CVE-75)

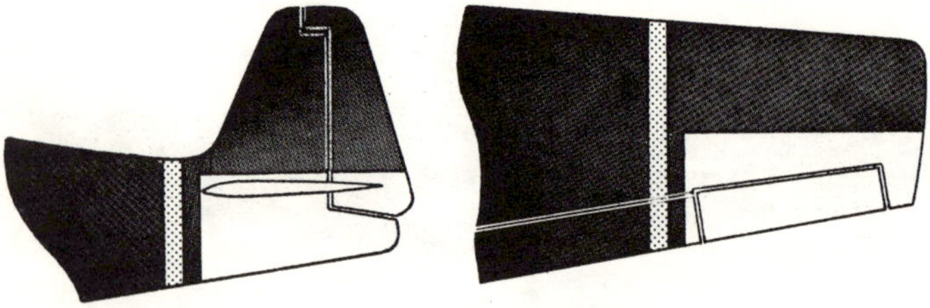

U.S.S. WAKE ISLAND (CVE-65)

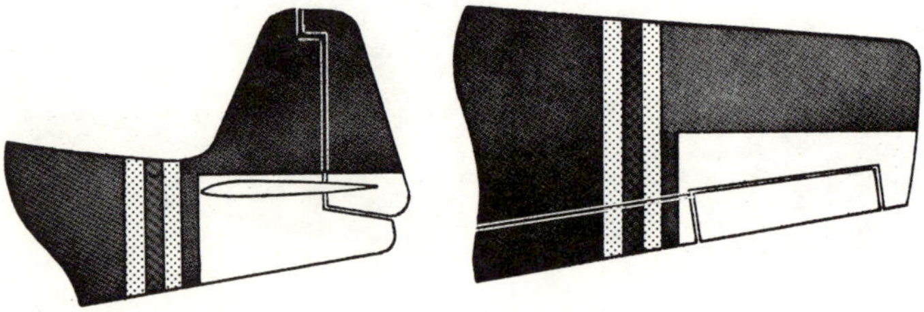

U.S.S. NEHENTA BAY (CVE-74)

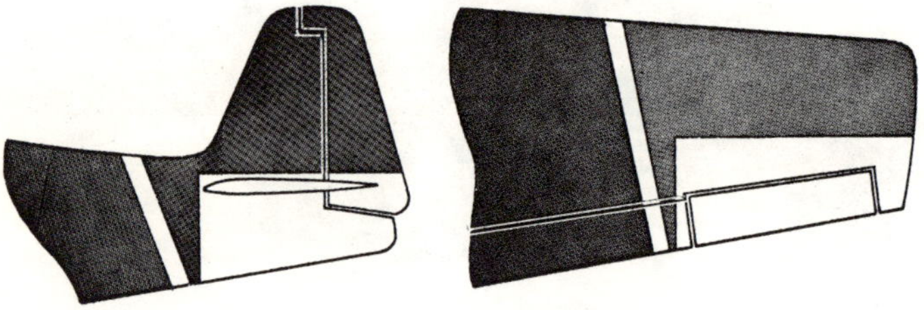

U.S.S. SHAMROCK BAY (CVE-84)

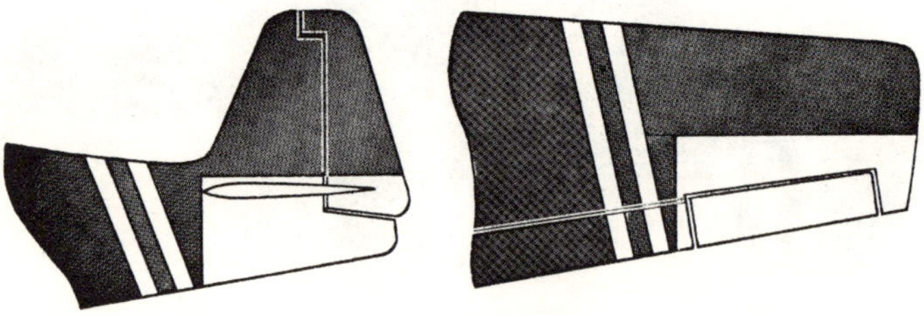

U.S.S. ANZIO (CVE-57)

CARRIER DIVISION TWENTY-FOUR

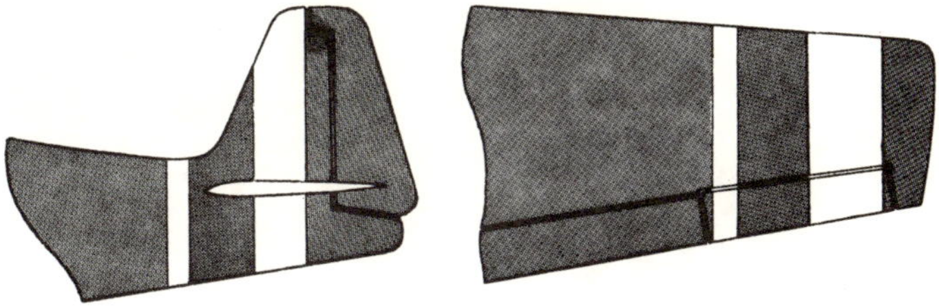

U.S.S. MARCUS ISLAND (CVE-77)

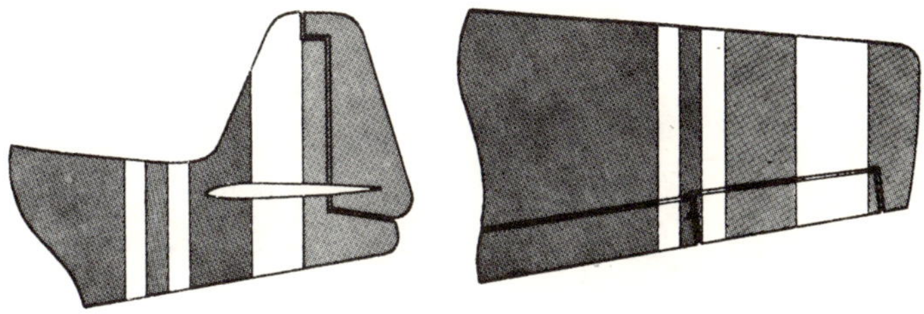

U.S.S. SHIPLEY BAY (CVE-85)

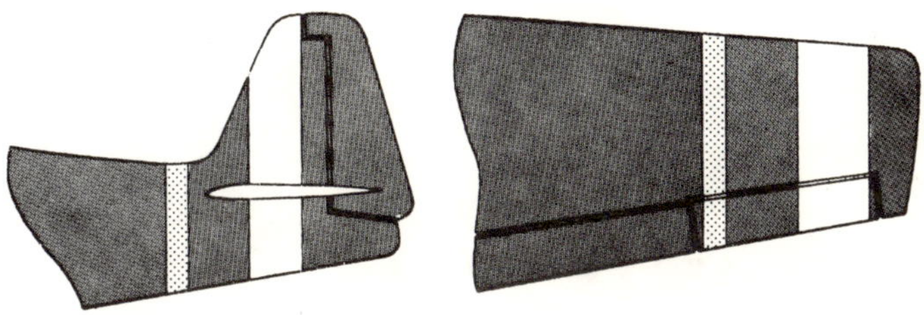

U.S.S. SAVO ISLAND (CVE-78)

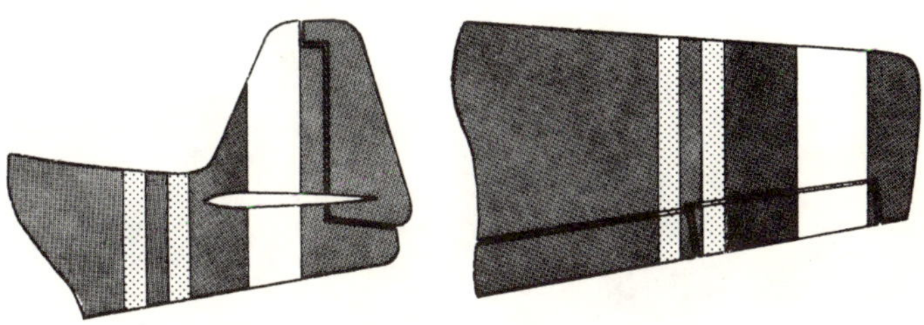

U.S.S. STEAMER BAY (CVE-87)

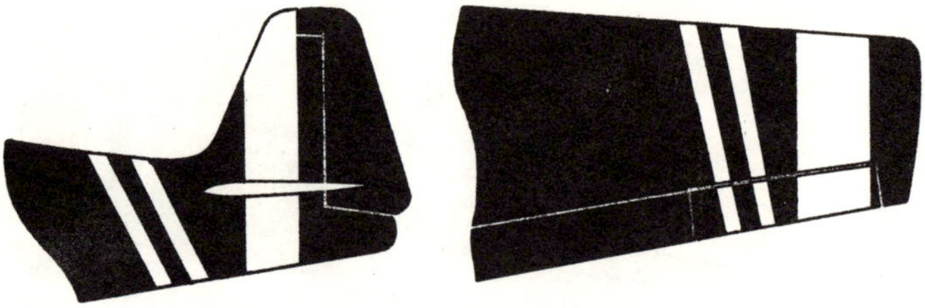

U.S.S. KASAAN BAY (CVE-69)

CARRIER DIVISION TWENTY-FIVE

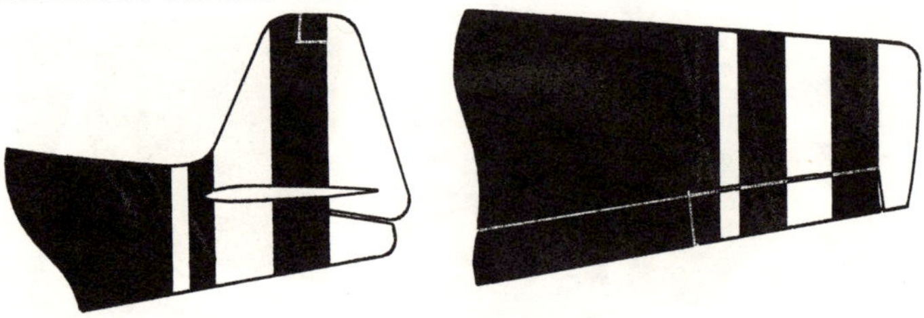

U.S.S. SAGINAW BAY (CVE-82)

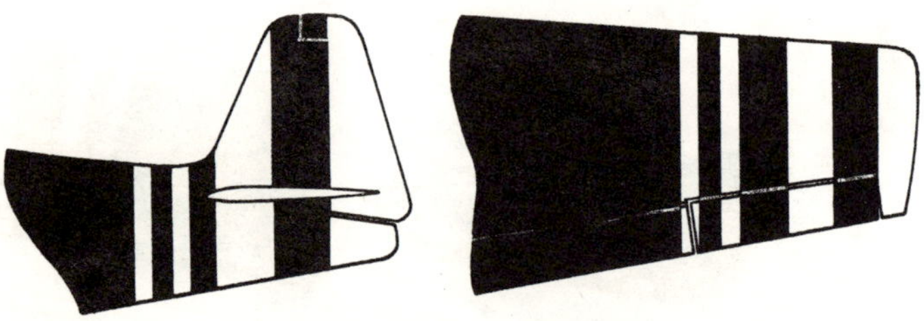

U.S.S. NATOMA BAY (CVE-62)

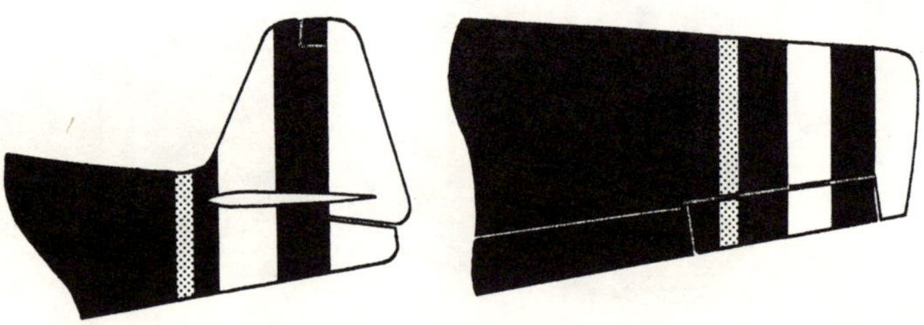

U.S.S. PETROF BAY (CVE-80)

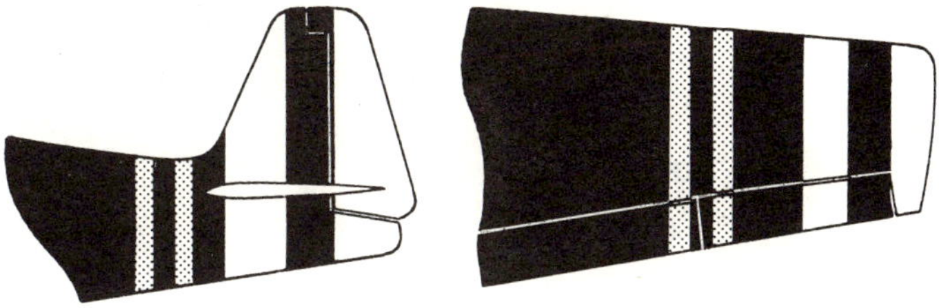

U.S.S. RUDYERD BAY (CVE-81)

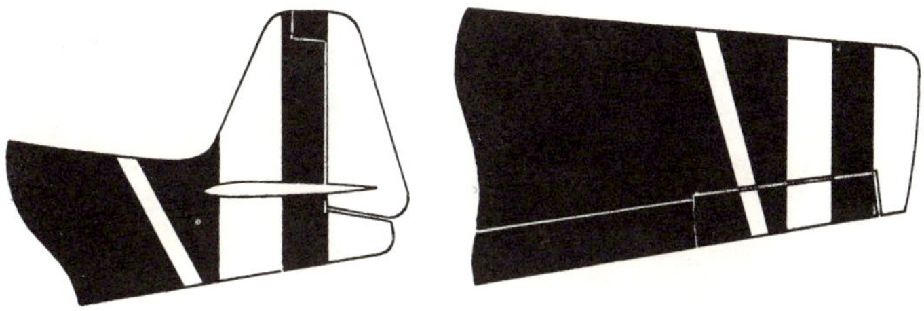

U.S.S. SARGENT BAY (CVE-83)

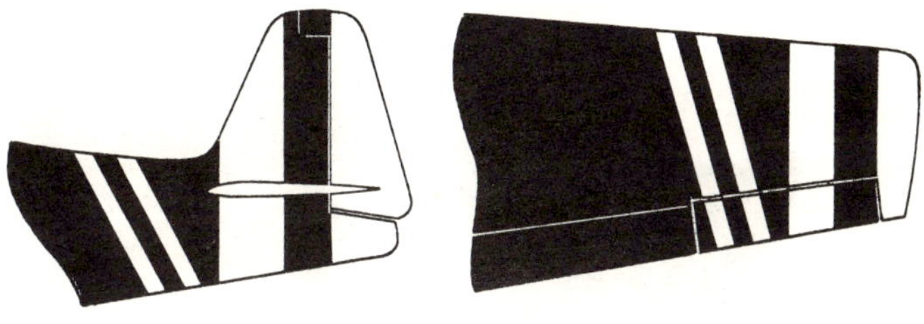

U.S.S. TULAGI (CVE-72)

CARRIER DIVISION TWENTY-SIX

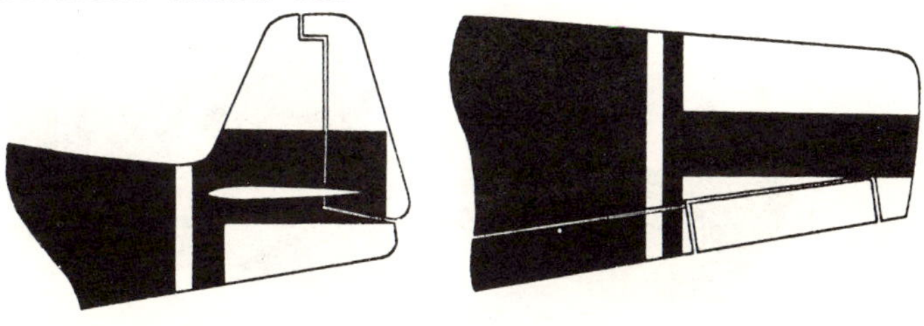

U.S.S. MAKIN ISLAND (CVE-93)

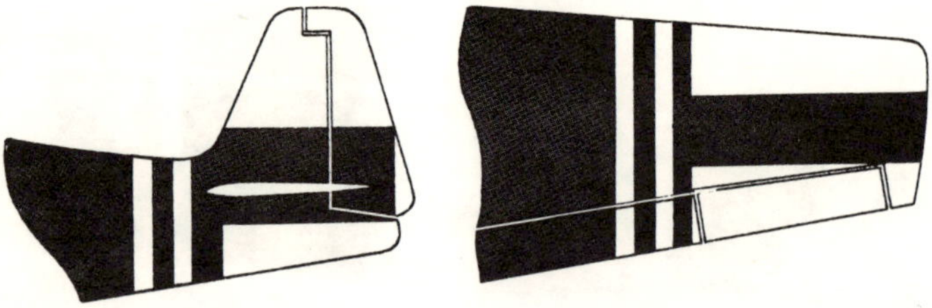

U.S.S. FANSHAW BAY (CVE-70)

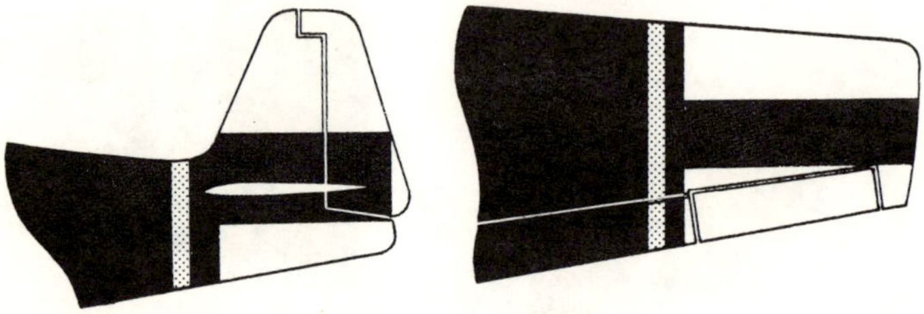

U.S.S. MANILA BAY (CVE-61)

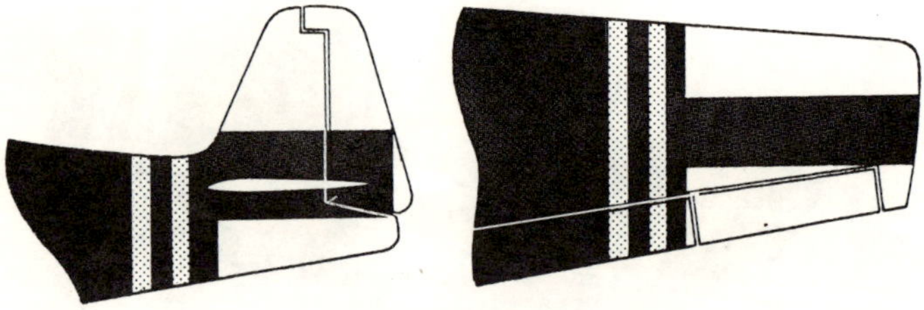

U.S.S. LUNGA POINT (CVE-94)

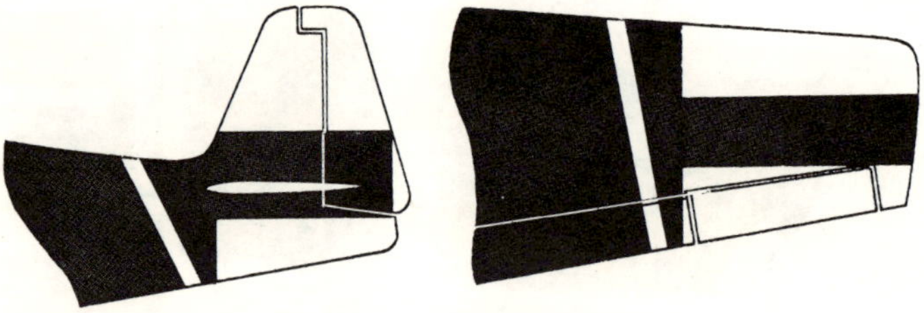

U.S.S. SALAMAUA (CVE-96)

CARRIER DIVISION TWENTY-SEVEN

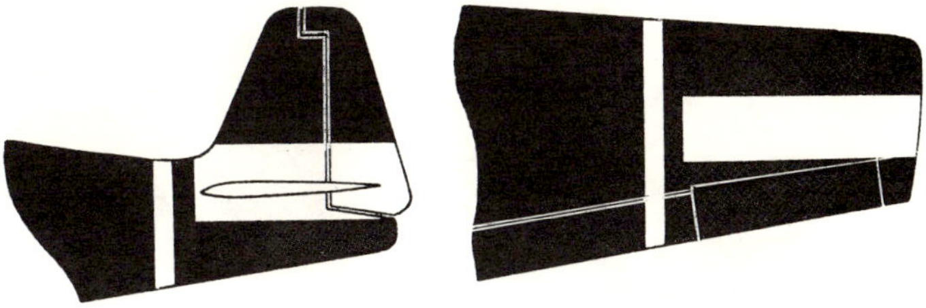

U.S.S. BLOCK ISLAND (CVE-106)

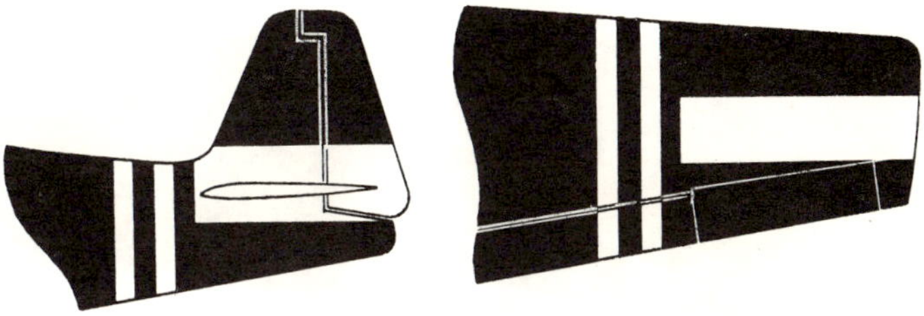

U.S.S. GILBERT ISLAND (CVE-107)

U.S.S. KULA GULF (CVE-108)

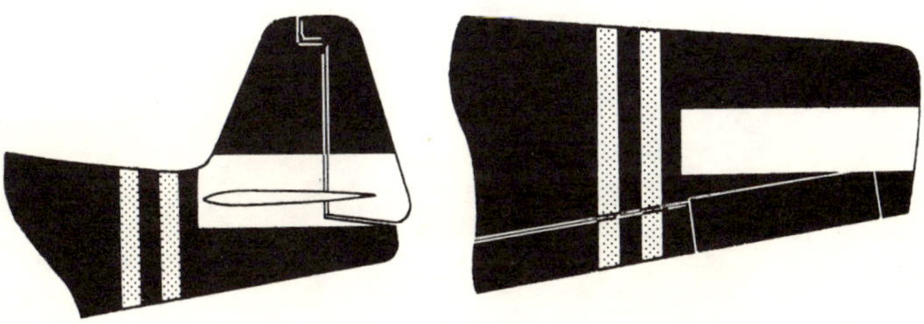

U.S.S. CAPE GLOUCESTER (CVE-109)

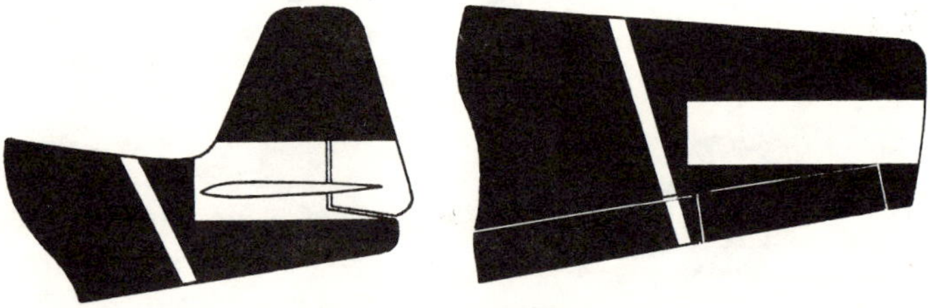

U.S.S. VELLA GULF (CVE-111)

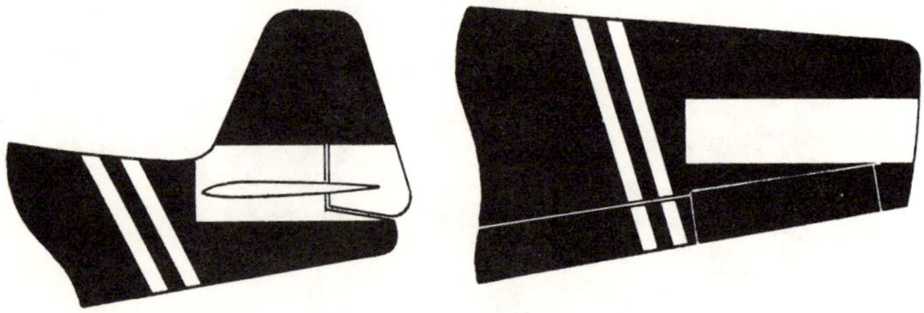

U.S.S. SIBONEY (CVE-112)

These markings remained in use until shortly after VJ Day, when they were discontinued.

1947

White ID letters were applied to all aircraft operating off escort carriers. The application was the same as for fleet carriers in the previous chapter, except that all escort carriers were assigned double letters. A few examples of these letters are shown below:

U.S.S. Badoeng Strait (CVE-116) .... CVEG-1 .. BS
U.S.S. Mindoro (CVE-120) .... CVEG-3 .. MI
U.S.S. Siboney (CVE-112) .... SI
U.S.S. Sicily (CVE-118) .... CVEG-2 .. SL
U.S.S. Saidor (CVE-117) .... SR
U.S.S. Bairoko (CVE-115) .... BA

8 MISCELLANEOUS MARKINGS

A variety of special markings were adopted for use from the 1920's to the mid-1940's; some of them saw service only briefly while others remained in use for many years. These markings are covered individually in the following pages, and include the approximate period of usage.

"Shellback" Turtle Insignia

The Winged Turtle was awarded to aircraft which had been flown across the equator. It was normally applied to both sides of the verticle tail, but some photos show it was also carried below and forward of the pilot's cockpit. The marking was applied in black.

Lt. W.V. Saunders and Ens. J.R. Lee were the first pilots to receive the insignia while carrying out reconnaissance of the Galapagos Islands in March, 1931.

The marking remained in use until the late 1930's when it was discontinued because flying across the equator was no longer an unusual achievement.

Unidentified Pennent

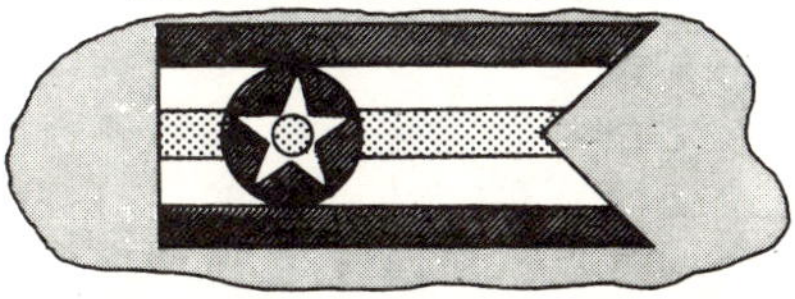

This marking is from a photograph of a Naval Aircraft

Factory PN-7 flying boat taken in 1925. The aircraft belonged to Seaplane Scouting Squadron One and the marking was carried on the forward fuselage near the bow. The purpose of this marking I have been unable to determine.

Excellence "E"

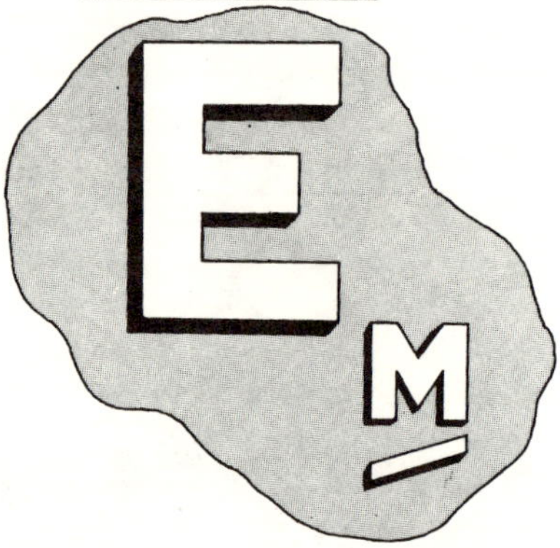

The E was awarded to pilots who qualified in the annual Individual Battle Practice and was carried on both sides of the fuselage forward of the pilot's cockpit. If only parts of the I.B.P. were qualified, then a smaller white M or a black B were applied behind the large E. The M indicated machine gunnery qualification and the B indicated bombing qualification. A slanting stripe, 'hash mark', was applied below the letters for each successive year that the pilot qualified in the competitions.

Reference has been found that indicates a gold E was applied when the pilot had qualified for five successive years; I have not found anything showing that this practice was used. Curtiss SOC-1 (#9858), 3-CS-7, carried four hash marks in 1938, which was a rare achievement.

The E was used from the late 1920's until World War II.

Reference has also been found that mentions a Communica--tions "C", which was applied in the same manner as the E. I have not yet found any example of this being actually applied.

Gunnery Trophy Pennent

This pennent was awarded to one squadron of each type for general excellence in gunnery competitions and was usually carried on the forward fuselage ahead of, or below, the pilot's cockpit. The G on the pennent was occassionally applied in white instead of black.

The pennent was used from the late 1920's to the mid or late 1930's.

Landing Assistance Stripes

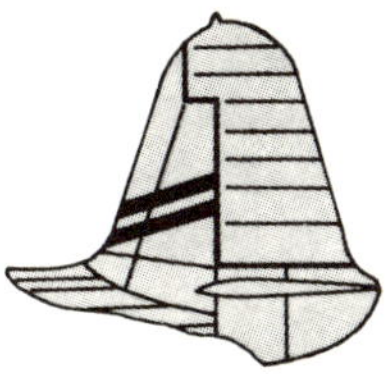 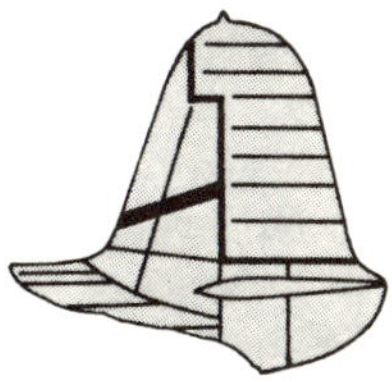

These stripes were applied to the left side of the vertical tail to aid the Landing Signal Officer in aligning the aircraft in proper landing attitude as it approached the carrier.

The stripes were applied in black on light colored or nat--ural metal tails and in white on dark colored or camouflaged tails. They were applied in either of the two methods shown in the above illustrations.

The stripes indicated as follows:

1) the double stripe system..... if the top stripe only was seen -- the pilot was coming in at too high an angle; if the top stripe and the portion between the stripes was seen -- the landing angle was OK; if both black stripes were visable -- the landing angle was too low.
2) the single stripe system..... indicated the angle was OK if

the stripe was just visable; too high if the stripe was not seen; too low if the stripe and tail color below was seen.

The stripe was used from the mid-1930's until the early 1940's.

Sharkmouth Marking

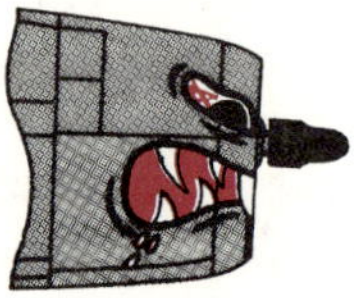 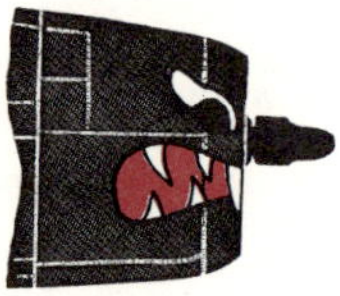

VF-27 aboard U.S.S. PRINCETON (CV - 23)
May - October 1944 October 1944

The markings for VF-27 were conceived while the squadron was training at Kahului N.A.S., Hawaii, during March and April 1944. Many designs were tried and the final one selected was the snarling mouth shown above. The bloodshot streaks did not show up well and were not included in the design on replacement aircraft. When the new sea blue over-all scheme came into use in October the drops of blood, accent marks, and the black out--line were also eliminated because they did not show up well.

Fighting Twenty-Seven's good luck at retaining their non--regulation markings ended on the 24th of October when the Princeton was sunk and nine surviving aircraft landed on other carriers. The Carrier Commanders took a dim view of this and within a few days all the snarling mouths were painted over.

VF-27 was the only unit aboard the U.S.S. Princeton to carry the snarling mouth on their aircraft. Aircraft operating from the Princeton did not carry any identification markings as was common to aircraft operating off other carriers.

Convair-Stinson OY-1 spotter aircraft operating on Okinawa in early 1945 also sported a sharkmouth marking. These aircraft were finished in olive drab with light grey wing under surfaces.

Convair-Stinson OY-1's on Okinawa in early 1945

These are the only two examples I have found of shark-mouth markings being used on U.S. Navy aircraft.

Colored Cowling

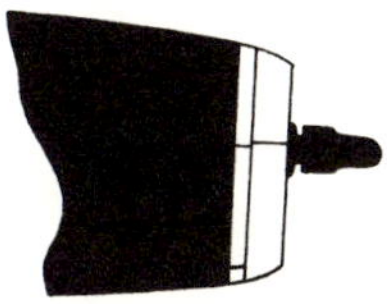

Fighter aircraft from carriers in Task Group 58.1 had the forward portion of their cowlings painted white for a short period in March, 1945, during the Okinawa campaign. The carriers and squadrons using the white forward cowling are shown below.

U.S.S. Wasp (CV-18)	VF-86
U.S.S. Bennington (CV-20)	VF-82
U.S.S. Belleau Wood (CVL-24)	VF-30
U.S.S. San Jacinto (CVL-30)	VF-45
U.S.S. Hornet (CV-12)	VF-17

Aircraft with Air Group 84, aboard U.S.S. Bunker Hill, sported yellow noses during the period 16 - 23 February 1945. The yellow nose may also have been used around mid-March '45.

Aircraft Numbers

Aircraft of VF-19 did not carry the 'hollow triangle' ID marking used by other aircraft operating from the U.S.S. Lex--ington in 1944. It was felt that the placing of the aircraft number behind the canopy was adequate identification for their Hellcats.

VF-19

VF-1

During the period May - July, 1944, VF-1 aboard the U.S.S. Yorktown carried the aircraft number on the upper right and lower left wing tips, in addition to the fuselage and tail locations. Also during this period VF-1 carried the letter 'K' (illustrated on page 68) on the rudder of their aircraft. I have not been able to establish the significance of the letter 'K' at this time.

The unusual location for the aircraft number, as used by VF-19, was utilized by two other squadrons in conjunction with the carrier ID tail markings. These two are shown below:

VF-53 aboard U.S.S. Saratoga (CV-3) early 1945
VF(N)-33 aboard U.S.S. Sangamon (CVE-26) early 1945

War Maneuver Markings

1932 Wide horizontal stripes were applied to the tail sur-faces of all aircraft taking part in the maneuvers. The color of the tail stripes was left to the dis-cretion of the Force Commanders.

1941 Large red crosses were applied in six locations, normally outboard of the National Insignia, on all aircraft participating in the joint Army-Navy maneu-vers.

Dive Flaps

During the summer of 1944 it was authorized that the interior of dive flaps be painted glossy insignia red.

Pilot's Assigned Aircraft

Pilots of Navy squadrons were assigned specific air-craft and their names were often painted on the cockpit rim.

A pilot of a land based squadron probably flew his own assigned aircraft whenever it was available for operations. However, pilots with carrier based squadrons seldom flew their assigned aircraft. This was due mainly to the fact that aircraft returning from a mission did not land in the order they were launched. Also, when the aircraft were spotted for takeoff, it was practically impossible to spot the aircraft in the order that the pilots were listed to takeoff.

During World War Two the assigned aircraft served as a scoreboard and victory markings were added as the assigned pilot accumulated them, regardless of what plane he was flying when he made a 'kill'. Any photographs of a pilot would then normally be taken of him seated in his assigned aircraft.

Aircraft operated from the U.S.S. Cabot carried victory markings indicating the number of 'kills' made in the aircraft itself, regardless of who the pilot was that made the 'kill'. This practice was intended to raise the morale of the crew servicing that particular aircraft; it probably did ----- because the U.S.S. Cabot had the lowest aircraft unservice-ability rate during W.W. II.

Personal Markings and Artwork

The application of personal markings and artwork to air-craft was frowned upon by the Navy, but there were a few cases of this being done. In all examples which I have found the markings were generally quite small and relativally simple. These markings were usually carried on the forward fuselage in front of and slightly below the pilot's cockpit.

VICTORY MARKINGS

Style A

Style A alternate

Style B

Style A ... This appears to have been the most common style used during World War II, and was occasionally applied with the long rays pointing to the rear. Fighting Squadrons carried the marking below the pilots cockpit on both sides of the fuselage.

Style A alternate ... This was used by Navy ace Lt. E.H. (Butch) O'Hare. It may have been a personal style or a less common version of style A.

Style B ... This style of victory marking was carried by aircraft #12 of VS-2 operating from the U.S.S. Lex--ington during Battle of Coral Sea in May, 1942, and piloted by Ens. J.A. Leppla. Four markings were carried below the pilot's cockpit and three below radio operator/rear gunner's position on the port side of the aircraft.

This style of victory marking was also carried by aircraft #12 of VF-15 aboard the U.S.S. Essex in 1943.

Victory markings began to appear in early 1942, but this practice did not become widespread as it was against regulations. Navy pilots applied their markings from top to bottom, rather than front to rear as was the practice in the A.A.F.

MISSION MARKINGS

Mission markings, in the form of a small white bomb, were used by some Torpedo and Bombing Squadrons. They were applied in rows, front to rear, below the pilot's cockpit on the port side of the aircraft.

SQUADRON INSIGNIA

Individual squadron insignias came into being during World War I when some units began to display them as a form of squadron pride. These markings were in general use by most Navy squadrons in the late 1920's and remained in use until the outbreak of World War II. The individual squadron insignia usually portrayed the squadron's basic mission, although some portrayed comic or mythological characters.

The insignia was usually large and generally carried ahead of the squadron mission coding, below the cockpit area on both sides of the fuselage. There were cases where some aircraft carried the insignia forward of the cockpit area on both sides of the fuselage.

Patrol Squadrons originally carried the insignia on the forward fuselage, but this was later standardized about mid--fuselage on both sides of the aircraft.

In mid-1942 squadron insignias began to appear on carrier based aircraft again, but in a much smaller size. The size of the insignia was not to exceed an area bounded by a 6" square.

The insignia was now carried slightly below, and forward of the cockpit area on both sides of the fuselage.

Listed below are three good references which illustrate many of the squadron insignias.

1) The National Geographic Magazine, June 1943 (color)
2) U.S. Navy Aircraft 1921-1941, Larkins (black & white)
3) U.S. Naval Institute Proceedings, vol. 59, January to June 1943 (black & white)

UNIDENTIFIED MARKING

Grumman F6F-5N Hellcat

The above illustrated marking was carried by aircraft #4 of VF(N)-33 aboard the U.S.S. Sangamon (CVE-26) in March, 1945, while operating off Okinawa. The significance of this marking, and whether it was carried by other squadron aircraft, remains undetermined. I assume that this marking was applied to both sides of the fuselage, but have found nothing to con--firm this assumption.

BOMB COLORS

It appears as though live bombs were painted yellow until mid-1941, when the color was changed to olive drab with yellow bands around the front and rear of the bomb body.

Practice bombs during World War II were light blue; this color was probably used prior to World War II for practice bombs also.

GRUMMAN F3F-1 (0253) PETER M. BOWERS' COLLECTION

Fourth Section Leader's aircraft, VF-4, carrying the 'Winged Turtle' marking indicating that the aircraft has flown over the equator. Note also the 'Red Rippers' squadron insignia on the forward fuselage, and the pilot's name, Lt. Hamilton, on the lower canopy frame. VF-4 was based aboard the U.S.S. Ranger and had solid willow green tail color.

NAVAL AIRCRAFT FACTORY PN-7 (A-6616) U.S. NAVY (3520)

Aircraft from Seaplane Patrol Squadron One, photographed on 7 April 1925. The purpose of the pennant marking on the bow remains undetermined. Note the squadron insignia just ahead of the squadron coding; this appears to be two 'Man-in-the-Moon' crescents with a large number 1 between them. There is a large eye above the '1' with the letters 'V' and 'S' on the left and right of the '1'. The word 'Scouting' is printed inside the upper edge of the circle.

CURTISS F6C-3 (A-7144) PETER M. BOWERS' COLLECTION

The 'Red Rippers', VB-1, operated their aircraft on floats for a short period during 1928. Note the squadron insignia and the 'E', for excellence in bombing, on the forward fuselage. The aircraft has a red spinner and solid lemon yellow tail surfaces.

VOUGHT SU-3 (9125) U.S. NAVY (450877)

Fighting Squadron One's utility plane photographed at San Diego on 25 January 1934. Squadron utility planes carried the squadron insignia and tail color, but did not use any section mark--ings. Note the Gunnery Pennant below the pilot's cockpit; this pennant was awarded to the top scoring squadron, in each squadron classification, during the Annual Battle Practice.

DOUGLAS SBD-3 DAUNTLESS (4628) PETER M. BOWERS' COLLECTION

Aircraft of Scouting Squadron Two with a non-standard light green camouflage on the upper surfaces at N.A.S. Oakland, 30 September 1941. Notice the single landing assistance stripe on the vertical tail, and the small '9' near the top of the vertical tail. The standard color scheme at this time was light grey over-all; the above color scheme may have been an early trial camouflage scheme.

CURTISS SBC-3 HELLDIVER SMITHSONIAN INSTITUTE (A 281-8)

This photograph shows the large red crosses applied to aircraft participating in the joint Army - Navy maneuvers in 1941. The crosses, normally applied outboard of the National Insignia, are carried inboard of the insignia because of its application near the wingtip, and the rear of the fuselage.

GRUMMAN F6F-3 NAVY DEPT., NATIONAL ARCHIVES (80-G-651629)

Personnel of VF-27 pose in front of a 'snarling' Hellcat at Kahului N.A.S., Maui, Hawaii, prior to embarking aboard the U.S.S. Princeton on 29 May 1944. The snarling mouth design was hand painted on all twenty-four of VF-27's Hellcats by the pilots; the 'snarling mouth' design was not carried by any other aircraft operating from the U.S.S. Princeton.

GRUMMAN F6F-5 NAVY DEPT., NATIONAL ARCHIVES (80-G-258158)

Cdr. C.A. Brown landing his damaged Hellcat aboard the U.S.S. Essex on 24 October 1944. The Princeton, VF-27's own ship, is burning in the background and later had to be sunk by a torpedo from the U.S.S. Reno. Note aircraft name, 'Paper Doll', and three victory markings below cockpit. This is one of two known photographs of VF-27 during combat.

DOUGLAS TBD-1 NAVY DEPT., NATIONAL ARCHIVES (80-G-66074)

LCdr. L.E. Massey, Executive Officer of VT-6, based aboard the U.S.S. Enterprise. During the raid on Kwajalein, 1 February 1942, he torpedoed a Japanese vessel and applied this unusual victory marking.

Victory markings, in the form of Japanese flags, began to appear in early 1942; the above form is one of the earliest styles used. The most commonly used victory marking is shown on the next page.

GRUMMAN F6F-5 (70143) NAVY DEPT., NATIONAL ARCHIVES (80-G-258198)

Cdr. David McCampbell, the Navy's top scoring Ace, in the cockpit of 'Minsi III' showing 30 victory markings. Cdr. McCampbell was credited with 1½ victories on 13 September 1944, bringing his total victories to 21; his last victory, number 34, was obtained on the 11 November 1944.

'Minsi III' was applied in yellow letters below the cockpit on the starboard side of the aircraft.

McCampbell was the Air Group Commander of Air Group 15 aboard the U.S.S. Essex during the above period.

GRUMMAN F4F-3 NAVY DEPT., NATIONAL ARCHIVES (80-G-64833)

Lt. E.H. (Butch) O'Hare photographed at Kahului N.A.S., Maui, Hawaii, in April 1942. Note the reduced size of VF-3's 'Felix The Cat' insignia, the name "Lt. O'Hare" below the cockpit, and the five (style A alternate) victory markings below the cockpit.

Lt. Cdr. O'Hare, killed in action on 25 November 1943, with 12 victories to his credit is in 22nd place on the list of top scoring Navy aces.

GRUMMAN TBF-1E AVENGER NAVY DEPT., NATIONAL ARCHIVES (80-G-366850)

Aircraft from VT-34, based aboard U.S.S. Monterey, carrying I.D. letters adopted in late July 1945. Note the 21 torpedo mission markings below the pilot's cockpit, and the personal artwork on the forward fuselage. Artwork was very rare on Navy air--craft; most personal markings were in the form of a name, applied in small letters.

GRUMMAN F6F-5 HELLCAT PETER M. BOWERS' COLLECTIC

Cdr. P. Maxwell, Commanding Officer of VBF-87, makes the U.S.S. TICONDEROGA's 10,000th landir in the 10,000th Hellcat, June 1945. The TICONDEROGA's white triangle marking is carried on both sides of the vertical tail, and on the upper right and lower left wing tips.

GRUMMAN F6F-5E U.S. NAVY

Aircraft of VF(N)-33 aboard the U.S.S. Sangamon (CVE-26) during operations off Okinawa in late March 1945. Note the aircraft number carried behind the cockpit, the step marker on the fuselage side and the marking over the National Insignia. The purpose of this marking over the insignia is unknown.

9 SPECIAL PURPOSE AIRCRAFT

COMMAND AIRCRAFT

Command aircraft were originally finished in standard color scheme and markings, but during the late 1920's many changes were made to denote aircraft assigned to officers of various command levels. By 1932-33 a system, similar to that used on Admirals' barges aboard ship, had been well standard--ized for use on command aircraft. The general color schemes for various command levels were as shown below:

COMMANDER-IN-CHIEF OF THE UNITED STATES FLEET (CinCUS)

a) Aircraft was over-all dark blue with silver wings.
b) 'U.S. FLEET' was carried in white letters on both sides of the rear fuselage.
c) National Insignia was carried as normal.

ASSISTANT SECRETARY OF THE NAVY FOR AERONAUTICS (AsstSecNav)

a) Aircraft was over-all dark blue with silver wings and horizontal tail surfaces. Silver trim was also applied to the fuselage of various aircraft assigned, usually in the form of lengthwise stripes.
b) 'U.S. NAVY' sometimes applied in white on rear fuselage.
c) National Insignia was carried as normal.

FLAG AIRCRAFT (Admiral rank)

a) Aircraft was over-all dark blue with silver wings and tail surfaces.
b) 'U.S. NAVY' carried in white on rear fuselage.
c) Command title, such as 'BATTLE FORCE', 'CRUISER SCOUTING FORCE', etc, was carried in white letters on the fuselage sides behind the cockpit area.
d) National Insignia was carried as normal.

COMMANDING OFFICER'S AIRCRAFT (Captain rank)

a) The aircraft was over-all silver with dark blue forward fuselage and landing gear. The blue of the fuselage ended about the trailing edge of the lower wing and then curved upward to a point behind the cockpit area.

b) 'U.S. NAVY' and National Insignia was carried as normal.

STAFF AIRCRAFT (aircraft not assigned to a unit commander as his personal aircraft)

a) Aircraft was over-all silver with standard markings and insignia.

b) The name of the Flag Unit, such as AIRCRAFT BATTLE FORCE, was carried in black on the fuselage sides behind the cockpit area.

Additional information on the above command aircraft:

a) The use of red, white, and blue rudder stripes was optional, but they were not usually applied.

b) The wheel pants on all land based aircraft had a silver horizontal stripe or bar.

c) Holders for rank cards were attached just below the cock--pit; these were normally mounted on the outboard side of the engine nacelles on twin engine aircraft.

d) Wing struts on bi-planes were painted the same color as the fuselage.

AIR GROUP COMMANDER'S AIRCRAFT (adopted 1939 or 1940)

a) Standard color scheme, markings, and National Insignia were used.

b) The carrier's assigned tail color was applied as normal to the tail and also in the form of a diagonal stripe running from in front of the cockpit to the trailing edge of the wing fillet.

c) The title 'Air Group Commander', with the carrier's name underneath, was carried in black letters below the cock--pit on both sides of the fuselage.

Author's comments:

a) Yellow upper wing surface carried as normal on all command aircraft.

b) Command Colors discontinued when camouflage was introduced.

AMBULANCE AIRCRAFT

During the early 1920's a special color scheme was used on some ambulance aircraft. This color scheme consisted of the aircraft being painted white over-all, with the upper surfaces of both wings and the horizontal tail being painted red.

A red stripe, ending at the cowling, was painted down the full length of the fuselage. There was a break in the stripe below the cockpit and a red cross was applied to the fuselage sides in this area. The wording 'U.S. NAVY' was carried in white letters on the red stripe just ahead of the tail.

A large white circle, with a red cross superimposed, was carried on the upper wing surfaces in place of the National Insignia. The National Insignia on the under wing surfaces was replaced by a large red cross.

Rudder stripes were carried as normal on the tail.

Reference has been found which indicates that ambulance aircraft were painted yellow over-all during World War II.

RESCUE AIRCRAFT

In 1945 a new color scheme was introduced for aircraft whose primary role was air-sea rescue. These aircraft were finished in ASW Scheme II with chrome yellow being applied to approximately the last six feet of the wing tips, wing floats and supporting struts, and a three foot wide band around the fuselage just ahead of the tail. A six inch black border was applied to the yellow band and the wing tips. Some aircraft also had a large yellow rectangle on the nose with a black border and the word 'RESCUE' in black. Later some air--craft also carried yellow on the upper surface of the wing center section, ending at the outer edge of the engine nacelle, with a black border and the word 'RESCUE' in large black letters.

'RECALL' AIRCRAFT

During World War II some large training centers in the Gulf Coast area used recall planes because primary trainers did not have radio equipment for recall in an emergency. When these planes were sent up it meant that all students were to return to base at once.

These aircraft were completely painted with equally spaced stripes which contrasted with the basic color scheme of the aircraft.[1]

'OPERATION CROSSROADS' AIRCRAFT

Several Grumman F6F-3K drone aircraft were flown through the radio-active mushroom cloud over Bikini Atoll after the atomic bomb tests in July, 1946.

These aircraft were painted red over-all with the complete tail section being painted white. National Insignia as normal.

TARGET DRONES

During the late 1930's drone aircraft were painted yellow over-all. In October, 1944, the color scheme was changed to glossy red over-all. National Insignia was carried as normal and a large black drone number was carried on the fuselage sides, an example is D-5. This number was later moved to the vertical tail and the D was dropped.

1. I have not seen a photo of these aircraft, nor have I found a scheme giving the colors used. However, I assume that the color scheme was probably yellow over-all with black diagonal stripes, similar to R.C.A.F. target tugs.

TARGET TUG AIRCRAFT

Instructions were issued on 7 October 1944 which authorized all aircraft employed on target tug duties to be painted chrome yellow over-all.

A new color scheme was issued for target tug aircraft on 2 January 1947. This color scheme consisted of the entire aircraft being painted glossy sea blue, except for the wing and tail surfaces which were painted glossy chrome yellow. The rudder and two wide bands around the wings were glossy insignia red. All other markings were carried as normal.

TRAINING AIRCRAFT

By late 1917 it had become the trend to paint all trainers yellow over-all. The reason for this was to differentiate between trainers and regular service aircraft and to provide for maximum visibility.

A directive was issued on 15 October 1938 which instructed that trainers were to be painted chrome yellow over-all with aluminum landing gear or floats.

On 24 April 1942, two new color schemes were issued for training aircraft. Primary trainers were to be painted glossy chrome yellow over-all with light grey landing gear. Advanced trainers were to be painted glossy aircraft grey with glossy chrome yellow upper wing surfaces.

New color schemes were issued on 6 March 1944, instructing that primary trainers be painted glossy chrome yellow over-all, and advanced trainers were now painted aluminum over-all.

It appears as though some training squadrons or bases used colored tail markings. Examples are shown below:

1) Consolidated P2Y-2 from VN-4D8 based at N.A.S. Pensacola in 1941 had single horizontal stripe on rudder. Stripe appears to be red in color.
2) Instrument trainer of VN-2 at Corry Field, Florida, in 1938 had a solid red rudder.

3) Naval Aircraft Factory N3N-3 instrument trainer based at N.A.S. Pensacola in 1942 had a solid light blue rudder.

INSTRUMENT TRAINING AIRCRAFT

Special identification markings were issued for instru--ment training aircraft in 1937. These markings consisted of a three foot wide red band around the fuselage in front of the tail and around each wing, inboard of the National Insignia. The bands were only applied to the top wing on biplanes.

In February 1942 the color of the bands was changed from red to light green.

THE HEAVIER - THAN - AIR UNIT

The Heavier-Than-Air Unit, the Navy's most unique 'Squad--ron', was organized in May, 1932. The unit was designed for very long-range reconnaissance, and was to operate from two rigid airships. A hanger for six aircraft was built into the hull of each airship, with the aircraft being launched and re--covered while the parent airship was in flight.

Operations began from the U.S.S. AKRON in May, 1932, and continued until she crashed off the New Jersey coast in April, 1933. The unit transfered to the U.S.S. MACON when she was commissioned in June, and continued operations until she was lost off Point Sur, California in February, 1935.

From the summer of 1934 until the Macon was lost, the unit operated with their 250 lb. landing gear removed and a 30 gallon belly tank installed. This action increased the top speed from 176 to 190 mph, and extended the range from 175 to 225 miles.

The unit was composed of six sections, each comprised of one Curtiss F9C-2 Sparrowhawk. These aircraft were finished in standard color scheme, National Insignia, and rudder stripes. A distinctive color was applied to the complete cowling, engine face plate, wheel pants (if landing gear was installed), wing chevron (with point to the rear), and fuselage band of ALL aircraft. The name of the airship was painted in black across the fuselage band, on both sides of the fuselage. If the fuse--lage band was a dark color, then that portion of the airship name on the band was painted in white.

The bureau numbers of these six aircraft, and the color carried by each, are shown below:

9056	insignia red	9059	black
9057	white	9060	willow green
9058	true blue	9061	lemon yellow

Late in 1933 the aircraft were color coded as a unit; at this time the entire vertical and horizontal tail surfaces were painted blue-black. Also at this time the aircraft color was removed from the cowling and, if landing gear was installed, from the wheel pants also.

The unit insignia 'The Men on the Flying Trapeze', was carried on the right side of ALL aircraft in the unit. Aircraft numbers 9056 and 9057 also carried the insignia on the left side of the fuselage. The insignia, designed in early 1933, was carried until late 1934 when its use was discontinued.

The Sparrowhawks had attachment points for arresting hooks provided , but hooks were not installed when operating from the airships.

The last remaining Sparrowhawk (9056), serving at N.A.S. Hampton Roads, Virginia, was about to be scrapped in the summer of 1939. However, it was preserved and is now in the hands of the Smithsonian Institute in Washington, D.C., where it is awaiting restoration. (Author: late 1974 - 9056 is now restored)

Apparently, while serving at N.A.S. Hampton Roads, #9056 had the complete rear portion of the fuselage and tail surfaces painted insignia red.

RESERVE AIRCRAFT

The Naval Air Reserve was formed at Squantum, Mass., and Ft. Hamilton, N.Y., on 13 August 1923. The reserve was organized mainly as training squadrons and followed regular Navy color schemes and markings.

- - - - - 1928 - - - - -

Many bases were, by this time, using special markings for their reserve aircraft. Aircraft assigned to a base carried the name of the base on both sides of the fuselage, followed by a large individual aircraft number. An example of this marking is shown below:

NAVAL AIR STATION
SAND POINT
WASH. **2**

Squadrons that were assigned to a Naval District, carried squadron coding to identify the district to which they were assigned. An example is shown below:

VN-13RD11-2 — second aircraft, Reserve Training Squadron 13 assigned to District 11

- - - - - 1930 - - - - -

Reserve aircraft on some bases were now being marked to indicate their reserve status and location. The letters USNR, with the base name underneath, were carried on both sides of the fuselage. This was applied as shown below:

U.S.N.R.
N.A.S. ANACOSTIA

- - - early 1930's - - -

District markings were dropped from squadron coding and retained for administration purposes only. Normal section markings and colors were used by squadron aircraft, letter 'N' was used as squadron code letter. The name of the squadron's home base was applied under 'U.S. NAVY' on both sides of the rear fuselage. Some squadrons also began using solid tail colors about this time.

Aircraft that were being used by two squadrons carried a combined squadron code which identified the two squadrons and also the district they operated in. This was marked as shown below:

VN3-4RD3 Reserve Training Squadrons 3 and 4, District 3, based at New York.

Also during this period reserve units began to adopt individual squadron insignia. These insignias generally por--trayed the area in which the squadron was based. An example of these markings was the 'Golden Gater' insignia used by VN-14R based at N.R.A.B. Oakland, Calif. The insignia took its name from the Golden Gate Bridge at the entrance to San Francisco.

- - - - - 1937 - - - - -

Squadron coding was dropped from the aircraft markings about this time. The aircraft number only was now carried on the fuselage sides; section leaders carried their plane number centered on the fuselage band.

- - - late 1930's - - -

By the end of the 1930's the markings on reserve aircraft had been changed to indicate the base and the individual plane number assigned by the base.

- - - - - 1946 - - - - -

In November a wide yellow band was placed around the fuse--lage, just ahead of the tail.

- - - - - 1947 - - - - -

In January it was authorized to apply a 3 foot wide inter--national orange band around the rear fuselage of reserve air--craft. The blue background circle of the National Insignia was centered across the fuselage band on both sides of the fuselage.

Also at this time it was ordered that I.D. letters, 36 inches high, be applied in white to both sides of the vertical tail, also upper right and lower left wing tips. Two letters were normally applied, the first identified the base and the second identi--fied the aircraft type.

- - - - - late 1949 - - - - -

About this time the National Insignia was moved ahead of the fuselage band on both sides of the fuselage. The word 'NAVY', with the base name underneath, was applied in white across a break in the band on both sides of the fuselage.

BASE IDENTIFICATION LETTERS AS USED DURING 1947 - 1950 PERIOD

B - Atlanta	J - Jacksonville	T - Seattle
C - Columbus	K - Olathe	U - St. Louis
D - Dallas	L - Los Alamitos	V - Glenview
E - Minneapolis	M - Memphis	W - Willow Grove
F - Oakland	P - Denver	X - New Orleans
H - Miami	R - New York	Z - Squantum
I - Grosse Isle	S - Norfolk	

Aircraft type identification letters

A - Attack	P - Patrol	T - Trainer
F - Fighter	R - Transport	U - Utility

Author's comments:

Below are listed a few of the bases which used solid tail colors during the mid-1930's to 1940 period, and the color used.

NRAB Great Lakes, Ill.	red
NRAB Seattle, Wash.	green
NRAB New York, N.Y.	blue
NRAB Grosse Isle, Mich.	red
NRAB Long Beach, Calif.	black *

*(photos appear to indicate that green was used during the mid-1930's)

VOUGHT O2U-1 CORSAIR (A-792_) U.S. NAVY

Command aircraft of Admiral H.A. Wiley, Commander-in-Chief of the United States Fleet, operating from the U.S.S. Texas, Flagship of the U.S. Fleet, early 1928. Aircraft operated with floats when aboard ship and was changed to wheels for operation from a Naval Air Station when the ship was in port.

CURTISS SOC-3 SEAGULL (1079) PETER M. BOWERS' COLLECTION

Flagplane of the Commander-in-Chief of the United States Fleet during the period June 1938 - late 1940. Aircraft is finished in the blue and silver color scheme as adopted in the early 1930's. Curtiss SOC's were used extensively as Flagplanes by high ranking officers during the late 1930's and early 1940's.

LOCKHEED XRO-1 'ALTAIR' (9054) U.S. NAVY

Command aircraft used by David S. Ingalls, Assistant Secretary of the Navy for Aeronautics, during the period October 1931 - June 1, 1932. The XRO-1 was the Navy's first landplane to have fully retractable landing gear. The blue and silver color scheme was adopted for use in early 1930; silver trim was applied to the fuselage of each aircraft assigned to AsstSecNav. (David S. Ingalls, with 5 kills, was the only U.S. Navy ace of World War I.)

CURTISS SOC-1 (9877) SMITHSONIAN INSTITUTE (A 53262)

Flagplane of Commander Cruisers, Scouting Force, based aboard the U.S.S. Indianapolis during the period early 1936 - 1940. Note the rank card holder below the cockpit area, and the stripe on the wheel pants.

CURTISS SBC-3 HELLDIVER (0527) SMITHSONIAN INSTITUTE (A 53264)

Aircraft assigned to the Air Group Commander aboard U.S.S. Yorktown in 1940. The diagonal stripe on the fuselage of AGC's aircraft was the same color as the assigned carrier tail color, in this case it was red. The use of the AGC's stripe was discontinued when camou--flage was introduced in October 1941. During the early part of 1942 the letters 'GC' were applied to the upper wing surfaces to identify the Air Group Commander's aircraft. In June 1942, the Air Group number was applied to the fuselage sides of the AGC's aircraft, such as '2', which identified the Lexington Air Group Commander.

GRUMMAN F6F-3 NAVY DEPT., NATIONAL ARCHIVES (80-G-474470)

Air Group Commander's aircraft, Air Group 5, aboard the U.S.S. Yorktown in early September 1943. With the adoption of carrier I.D. markings, the '00' number was accepted as identification for the Air Group Commander's plane. In January, 1947, a new aircraft numbering system was brought into use and the identifying number for the Air Group Commander was changed to '100'. Note the small 'oo' under the tail, Yorktown's slanting stripe tail marking, and the camouflage application to fuselage sides.

de HAVILLAND DH-4B (A-6125) U.S. NAVY (644480)

Special ambulance modification operating with Aircraft Squadrons, Battle Fleet in December 1923. The upper surfaces of both wings and tail, fuselage stripe, and crosses are red; the rest of the aircraft is finished in white. The fuselage behind the pilot has been altered to enclose a litter.

MARTIN PBM-3S MARINER (6620) PETER M. BOWERS' COLLECTION

Rescue aircraft photographed in late 1946. Color scheme is white with grey upper surfaces and yellow rescue markings applied to the wing tips, floats, and band around the rear fuselage. Some aircraft also had a large yellow rectangle on the forward fuselage with a black border and the word 'RESCUE', in black letters, inside the rectangle.

MARTIN JM-1 MARAUDER U.S. NAVY

Aircraft of VJ-15 finished in the over-all chrome yellow color scheme adopted for use on target tug aircraft in October 1944.

DOUGLAS JD-1 (77211, 140344, 77173) PETER M. BOWERS' COLLECTION

Formation photo showing the colorful scheme adopted for use on target tug and drone control aircraft in January 1947. The fuselage and engine nacelles are sea blue, the wings and tail are chrome yellow, rudder and wing bands are insignia red. China Lake, Calif. during 1950's.

VOUGHT O3U-1 (8862) WILLIAM T. LARKINS

Instrument training aircraft attached to VF-3 during 1937. The red wing and fuselage bands were an air safety measure used to identify aircraft used for instrument training. The color of these bands was changed from red to light green in February, 1942.

Most squadrons had their own instrument trainer, usually an obselete type was used for this purpose. These aircraft carried squadron coding and, in many cases, the squadron tail color and insignia. They did not employ any section markings on aircraft numbered over 19.

U.S.S. MACON (ZRS-5) NAVY DEPT., NATIONAL ARCHIVES (80-G-428442)

The MACON prepares to recover her Curtiss F9C-2 fighters. Six aircraft were housed in a hanger built into her hull, and were used for long range reconnaissance and protection of the airship.

The Heavier-Than-Air Unit was transfered to the MACON when she was commissioned in June, 1933, and continued operations until she was lost in February, 1935.

CURTISS F9C-2 SPARROWHAWK (9057) U.S. NAVY

Lt. D.W. Harrigan hooking up MACON's number two aircraft for the first time, over New Egypt, N.J., on 7 July 1933. The aircraft's white section markings can be clearly seen in this photograph. Note also the unit crest designed by Lt. Harrigan.

CURTISS F9C-2 SPARROWHAWK (9056) PETER M. BOWERS' COLLECTION

Commander's aircraft, Heavier-Than-Air Unit, showing the first use of the blue-black tail color in November, 1933. The 'section' color has been removed from the cowling, but has not yet been removed from the wheel pants.

The colors for the 'Men on the Flying Trapeze' insignia are as follows ----
acrobats -- white, shorts -- red, vests -- yellow, background circle -- dark blue, inner border ring -- red, and outer border ring -- yellow.

CURTISS F9C-2 (9056) PETER M. BOWERS' COLLECTION

Unit Commander's aircraft, Macon number I, sporting the new style of blue-black tail in 1934. Notice that the 'section' color has been removed from the wheel pants, and that the unit insignia has also been removed. The aircraft also appears to have been recently repainted, and the wording 'U.S. NAVY' has been applied in larger letters.

CURTISS O2C-1 (8948) WILLIAM T. LARKINS' COLLECTION

Leader's aircraft, second section, Reserve Training Squadron 10, based at NRAB Great Lakes, Ill., in 1934. Tail color is solid red. Aircraft carried the standard color scheme and markings in use at this time.

GRUMMAN SF-1 (9473) WILLIAM T. LARKINS' COLLECTION

Second section leader's aircraft of the Naval Reserve Air Unit attached to N.A.S. Anacostia in 1937. Tail color appears to be solid blue. The use of solid colored tails or rudder stripes, by reserve bases and units, was optional during the mid-1930s period. The use of section markings without a squadron code aided to identify reserve aircraft.

CURTISS SBC-4 HELLDIVER (1816) PETER M. BOWERS' COLLECTION

Aircraft number one assigned to N.R.A.B. Oakland, photographed in late 1940, shows the new reserve aircraft markings adopted for use during the late 1930's. Note the 'Golden Gaters' insignia below the cockpit area, indicating their operation in the San Francisco area.

NORTH AMERICAN FJ-1 FURY (120352) WILLIAM T. LARKINS' COLLECTION

This aircraft from the Naval Air Reserve squadron at N.A.S. Oakland, California, carries the reserve markings as used during the late 1940's. Navy squadron VF-5A, the first jet-equipped Navy unit, took delivery of the Fury in November, 1947. Near the end of 1948 this squadron was re-numbered VF-51, and equipped with the new Grumman F9F-2; the Fury jets were then re-assigned to Naval Air Reserve units ashore.

GRUMMAN F8F-1 BEARCAT (95447) PETER M. BOWERS' COLLECTION

This aircraft, belonging to the Naval Air Reserve squadron at N.A.S. Akron, Ohio, illustrates the new markings adopted for use by reserve units in late 1949. The National Insignia was moved ahead of the fuselage band, and the base name was added under 'NAVY'.

10 DESIGNATION SYSTEMS

1911 - 1914

The first system adopted consisted of a letter - number group which combined the type of aircraft designator with a serial number. The letter designated the manufacturer and the number indicated the purchasing sequence of the aircraft.

The letter 'A' was assigned to aircraft built by Curtiss and the letter 'B' was given to machines purchased from the Wright Brothers. Thus, 'A-1' represented the Navy's first airplane and was built by Curtiss; 'B-3' was the third air--plane procurred and was built by the Wright Brothers.

By March 1914 there were five letters in use to identify different types of aircraft and manufacturers. These letters are shown below:

A -	Curtiss	Land and Hydro
B -	Wright	Land and Hydro
C -	Curtiss	Flying Boats
D -	Burgess	Flying Boats
E -	Curtiss	Amphibian

1914 - 1916

A new system was adopted on March 27, 1914, and all air--craft were re-designated as to type and sequence regardless of the manufacturer. Numbers were assigned to each aircraft in numerical sequence within each class, such as AH-1, AH-2, AB-1, AB-2, etc. The designations for each class of airplane are as follows:

AB - Flying Boats
AC - Convertible Airplanes
AH - Hydro Airplanes
AL - Land Airplanes
AX - Amphibians

This system was discontinued when a sequential serial number system was adopted in early 1917.

1917 - 1922

The Assigned Serial Number was called by different names over the years; such as Building Number, Designating Number, and since World War II it has become known as the Bureau Number. The number was assigned to the airframe and remained with the fuselage until the plane was scrapped or otherwise removed from service. The numbers, once used, were not re--assigned and are the only identification by which the air--crafts history can be traced.

Early in 1917 the combined type and number identification was discontinued and a sequential serial number system was adopted. The system began at 'A-51' and is still in use to--day with six digits. The letter 'A' was dropped from the number in 1930.

No specific aircraft designation system was used during this period. Aircraft were procured and operated under the manufacturer's name and model number, such as Curtiss H-12, Thomas-Morse SH-4, etc.

1922 - 1950 (System remained in use until 1962)

On March 29, 1922, a systematic aircraft designation system was adopted for use on all Naval aircraft. This system consisted of letters and numbers which identified aircraft as to manufacturer, aircraft type, and model sequence.[1]

When the system was first adopted for use the aircraft were identified by manufacturer. The first letter indicated the manufacturer, the next letter indicated the aircraft type,

1. This system applied to all naval aircraft with the following exceptions:

   1 - Civilian aircraft taken into service during World War II -- were operated under their original civilian model designations.

   2 - Captured enemy aircraft -- used their original designations with Navy serial numbers.

   3 - Army models, adopted in large numbers -- were assigned new Naval designations.

and the number represented the model number. Thus, MO-1 indicated Martin (M), first Observation type (O), and -1 was the initial design of the model. A recognizable change in the initial design would be the MO-2, MO-3, etc. An entirely new Martin Observation design would be the M2O-1.

On March 10, 1923, it was decided to identify aircraft by type rather than by manufacturer, so the type identifi--cation letter was moved to first place. Aircraft with the old designation retained it, as did subsequent re-orders of the same model. New models were designated under the new type system. Thus, UO-1 indicating Vought (U), first Obser--vation type (O), initial design remained as such; the next Vought Observation to be designed would now be the O2U-1 since aircraft were now identified by type.

The system became more detailed and eventually contained six parts as shown below:

```
X F 2 A - 2
  J 2 F - 2 A
-------------
5 1 4 2   3 6
```

1 - Type or Class
2 - Manufacturer
3 - Aircraft Configuration Sequence
4 - Manufacturer's Type Sequence
5 - Status or Class Prefix
6 - Special Purpose Suffix

Type or Class

A single letter to identify the aircraft type; in 1934 dual letters were adopted to identify aircraft with dual purpose such as Patrol Bomber, etc.

There were cases of conflict with type letters, such as 'T' for Transport being replaced by the letter 'R' because the letter 'T' was being used to indicate Torpedo. After World War II when the Torpedo designation was dropped, the letter 'T' was assigned to new Training aircraft; the old Training aircraft that remained in service continued to carry the letter 'N' until the end of their service.

Below are listed the type designation letters used from 1922 - 1950 with dates from initial assignment to termination of service for that type. Those designation letters in use beyond 1950 are indicated as 1950+ .

Letter	Classification	Service
A	Ambulance	1943
A	Attack (formerly BT)	1946 - 1950+
B	Bomber	1931 - 1943
BF	Bomber-Fighter	1934 - 1937
BT	Bomber-Torpedo	1942 - 1945
F	Fighter	1922 - 1950+
G	Transport, single engine	1939 - 1941
H	Helicopter (combination)	1943 - 1950+
H	Hospital (to A-1943)	1929 - 1931 & 1942
J	Transport	1928 - 1931
J	Utility	1931 - 1950+
JR	Utility Transport	1935 - 1950+
L	Glider (combination)	1941 - 1945
N	Trainer	1922 - 1950+
O	Observation	1922 - 1950+
OS	Observation-Scout	1935 - 1945
P	Patrol	1923 - 1950+
P	Pursuit	1923
PB	Patrol Bomber	1935 - 1950+
PTB	Patrol Torpedo-Bomber	1937
R	Racer	1922 - 1928
R	Transport (formerly J)	1931 - 1950+
S	Scout	1922 - 1946
SB	Scout-Bomber	1934 - 1946
SN	Scout-Trainer	1939 - 1950+
SO	Scout-Observation	1934 - 1946
T	Torpedo	1922 - 1935
T	Trainer	1948 - 1950+
T	Transport	1927 - 1930
TB	Torpedo-Bomber	1935 - 1946

TD	Target Drone	1942 - 1946
TS	Torpedo-Scout	1943
U	Unpiloted Drone	1946 - 1950+

Manufacturer

A single letter, normally the first letter of the manufacturer's name, was used to identify the manufacturer of the aircraft. The original manufacturer's designation letter was used regardless of what company actually built the aircraft, the only distinguishing feature between builders was the assigned serial numbers.

Problems arose and this proceedure was changed to indicate the actual builder rather than the original manufacturer; for example the Vought F4U Corsairs built by Goodyear were designated as FG Corsairs and the ones built by Brewster were designated as F3A Corsairs.

Some letters were re-assigned to new companys after the previous company had gone out of business. During World War II there were cases of more than one manufacturer using the same letter at the same time, such as the letter B indicating Boeing and Beech, etc.

Below are listed the letters assigned to companies from 1922 - 1950 including period of service from intial assignment until removal of company aircraft from service or change in company name. In cases where the manufacturer was under contract beyond 1950 this is indicated as 1950+ .

Letter	Manufacturer	Service
A	Aeromarine Plane and Motor Company	1922
A	Allied Aviation Corporation (Gliders)	1941 - 1943
A	Atlantic Aircraft Corporation (American Fokker)	1927 - 1930
A	Brewster Aeronautical Corp.	1935 - 1943
A	General Aviation Corporation (ex-Atlantic) (also J)	1930 - 1932

A	Noorduyn Aviation, Ltd (Canada)	1946
B	Beech Aircraft Company	1937 - 1945
B	Boeing Aircraft Company	1923 - 1950+
B	Budd Manufacturing Company	1942 - 1944
B	Aerial Engineering Corporation (Booth or Bee Line)	1922
BS	Blackburn Aeroplane & Motor Company (England)	1922
C	Cessna Aircraft Corporation	1943
C	Culver Aircraft Corporation	1943 - 1946
C	Curtiss Aeroplane & Motor Company	1922 - 1946
CH	Casper-Werke G.m.b.H. (Germany)	1922
D	Douglas Aircraft Corporation	1922 - 1950+
D	McDonnell Aircraft Corporation (later changed to H)	1942 - 1946
D	Radioplane Corporation (drones)	1943 - 1950+
D	Frankfort Sailplane Company (target drones)	1945 - 1946
DH	De Havilland Aircraft Company Ltd (England)	1927 - 1931
DW	Dayton-Wright Airplane Co.	1923
E	Bellanca Aircraft Corporation	1931 - 1937
E	Detroit Aircraft Corporation (changed to Great Lakes)	1928
E	Edo Aircraft Corporation	1943 - 1946
E	G. Elias & Brothers, Inc	1922 - 1924
E	Gould Aeronautical Corp. (gliders)	1942 - 1945
E	Hiller Aircraft Corporation (helicopters)	1948 - 1950+
E	Piper Aircraft Corporation (also P)	1941 - 1945
E	Pratt-Read (gliders)	1942 - 1945
F	Fairchild Aircraft, Ltd (Can.)	1942 - 1945

F	Royal Dutch Aircraft Manufac- -turing Works (Fokker)	1922
F	Grumman Aircraft Engineering Corporation	1931 - 1950+
G	Gallaudet Aircraft Corp.	1922
G	Bell Aircraft Corporation (Bell built Great Lakes)	1935 - 1936
G	Eberhart Aeroplane & Motor Company	1927 - 1928
G	Globe Aircraft Corporation (target drones)	1946 - 1950+
G	Goodyear Aircraft Corp.	1942 - 1950+
G	Great Lakes Aircraft Corp. (ex-Detroit)	1929 - 1935
G	A.G.A. Aviation Corporation (gliders)	1942
H	Hall-Aluminum Aircraft Corp. (into Consolidated)	1928 - 1940
H	Howard Aircraft Company	1941 - 1944
H	Huff, Daland & Company (changed to Keystone)	1922 - 1927
H	Stearman-Hammond Aircraft Corporation	1937 - 1938
H	Snead and Company (gliders)	1942
HP	Handley Page, Ltd (England)	1922
J	Berliner-Joyce Aircraft Corp.	1929 - 1935
J	General Aviation Corporation (ex-Atlantic)	1935
J	North American Aviation Corp. (successor to General)	1937 - 1950+
JL	Junkers-Larson Aircraft Corp.	1922
K	Fairchild Aircraft Corp. (ex-Kreider-Reisner)	1937 - 1942
K	Kaiser Cargo, Ltd (Fleetwings Div.)	1943 - 1945
K	Keystone Aircraft Corp. (ex-Huff-Daland)	1927 - 1930

K	Kinner Airplane & Motor Corp.	1935 - 1936
K	Kreider-Reisner Aircraft Co., Inc (to Fairchild)	1935
K	Martin, J.V.	1922 - 1924
K	Nash-Kelvinator Corporation	1942
L	Bell Aircraft Corporation (helicopters since 1946)	1939 - 1950+
L	Columbia Aircraft Corp.	1945
L	Grover-Loening, Inc	1932 - 1933
L	Langley Aviation Corporation	1942 - 1943
L	Loening Aeronautical Engineering Corp. (into Keystone)	1922 - 1932
L	L.W.F. Engineering Corp.	1922
M	Glenn L. Martin Company	1922 - 1950+
M	General Motors Corporation (Eastern Aircraft Div.)	1942 - 1945
N	Naval Aircraft Factory	1922 - 1945
O	Lockheed Aircraft Corp. (Plant B)	1931 - 1950
O	Viking Flying Boat Corp.	1929 - 1936
P	Piasecki Helicopter Corp. (ex- P-V Engineering)	1946 - 1950+
P	Piper Aircraft Corporation (gliders) (also E)	1942 - 1943
P	Pitcairn Autogyro Company	1931 - 1932
P	P-V Engineering Forum (became Piasecki, Vertol)	1944 - 1950+
P	Spartan Aircraft Company	1940 - 1941
PL	Parnell & Company (England)	1922
Q	Bristol Aeronautical Corp. (gliders)	1941 - 1943
Q	Fairchild Engine & Airplane Company	1928 - 1950+
Q	Stinson Aircraft Corp. (into Convair)	1934 - 1936
Q	Chas. Ward Hall, Inc	1926

R	Aeronca Aircraft Corp. (Army TG-5 glider)	1942
R	American Aviation Corp. (glider)	1942
R	Brunswick-Balke-Collender Corporation	1942 - 1943
R	Ford Motor Company	1927 - 1932
R	Interstate Aircraft & Engineering Corp. (drones)	1942 - 1950+
R	Maxson-Brewster Corp.	1939 - 1940
R	Radioplane Div, Northrop Corp. (target drones)	1943 - 1950+
R	Ryan Aeronautical Company	1941 - 1946
RO	Officine Ferroviarie Meridionali Romeo (Italian built Fokker C-V)	1933
S	Schweizer Aircraft Corp. (glider)	1941
S	Sikorsky Aviation Corp. (helicopters since 1943)	1928 - 1950+
S	Stearman Aircraft Company (to Boeing-Wichita, 1939)	1934 - 1945
S	Stout Engineering Laboratories	1922
S	Sperry Gyroscope Company (target drones)	1950
S	Supermarine (Spitfire as FS-1)	1943
T	New Standard Aircraft Corp.	1930 - 1934
T	The Northrop Corporation (to Douglas-El Segundo)	1933 - 1937
T	Taylorcraft Aviation Corp. (Army TG-6 glider)	1942
T	Northrop Aircraft, Inc (Army P-61 as F2T-1)	1944
T	Timm Aircraft Corporation	1941 - 1943
T	Thomas-Morse Aircraft Corp.	1922
U	Chance Vought Corporation	1922 - 1950+

V	Lockheed Aircraft Corp. (Vega, Plant A)	1942 - 1950+
V	Canadian Vickers, Ltd	1942 - 1945
V	Vultee Aircraft, Inc (into Consolidated)	1941
VK	Vickers, Ltd (England)	1922
W	Waco Aircraft Corporation (gliders since 1942)	1934 - 1945
W	Wright Aeronautical Corp.	1922 - 1926
W	Canadian Car & Foundry Co.	1942 - 1945
X	Cox-Klemin Aircraft Corp.	1922 - 1924
Y	Consolidated Aircraft Corp. (to Convair)	1926 - 1942
Y	Convair Aircraft Corp. (ex-Stinson, Consolidated)	1942 - 1950+
Z	Pennsylvania Aircraft Syndicate	1933 - 1934

Aircraft Configuration Sequence

The aircraft's initial configuration was identified with a -1 and any subsequent changes were identified with -2, -3, etc.

Type Sequence Number

This number identified different models of a similar aircraft type procured from the same manufacturer and also indicated the procurement sequence. An example is FF-1 (Grumman's first fighter), F2F-1 (Grumman's second fighter model), F3F-1 (Grumman's third fighter model), etc.

Status or Class Prefix

In 1927 the Status Prefix letter 'X' was added to the aircraft designation to indicate aircraft of prototype or experimental status.

Immediately after World War II, for a short period, the letter 'N' was applied to first-line aircraft assigned to Reserve Training Squadrons.

In 1950 the letter 'Y' was added to indicate prototype

aircraft that had advanced beyond the straight experimental stage. This was similar to the Army and Air Force service test status.

During World War II the letters 'H' (helicopters) and 'L' (gliders) were used with regular designations, such as HO2S-1 (Sikorsky's second observation helicopter), etc.

Special Purpose Suffix

Prior to World War II these letters were used sparingly to indicate minor changes in the aircraft configuration not warranting a change in the aircraft dash number, such as TBD-1A identifying the seaplane test version of the Douglas TBD-1 Devastator.

During World War II and the late 1940's these letters came into wide usage to identify aircraft used for purposes other than their designated mission, the addition of special equipment, etc. In some cases the aircraft was further mod--ified producing a second sub-variant of the same type; such aircraft were identified by a number after the suffix letter, such as DH-4B1. Confusion resulted occasionally due to mult--iple use of the same letter to indicate different versions of various aircraft.

Below is a list of suffix letters used during the 1922 to 1950 period:

A	Miscellaneous modification
A	Armament on a normally unarmed airplane
A	Arrester gear on normally non-carrier aircraft
A	Built for or obtained from Army
A	Amphibious version
A	Land-based version of carrier aircraft
B	Miscellaneous modification
B	Special armament
B	British (lend-lease) version
C	Carrier version of non-carrier aircraft
C	Reinforced for catapulting (this was the first suffix letter adopted for use)

C	Cannon armament
C	Navy equivalent of Army C-series aircraft
D	Drop tanks
D	Drone control
D	Navy equivalent of Army D-series aircraft
D	Special search
E	Special electronic gear
F	Flagship conversion
G	Search and rescue
G	Gun on a normally unarmed aircraft (later replaced by the letter A)
G	Navy equivalent of Army G-series aircraft
H	Hospital conversion
H	Navy equivalent of Army H-series aircraft
J	Special weather equipment
J	Navy equivalent of Army J-series aircraft
K	Target drone version
L	Winterized
M	Weather reconnaissance
N	Night operating aircraft
N	All-weather operation
NA	Day attack converted to night operating
NL	Night operating - winterized
P	Photographic
Q	Electronic countermeasures
R	Transport version
S	Landplane converted to seaplane
T	Trainer version
U	Utility
W	Early warning
Z	Administration transport

Suffix letters were also used by some manufacturers prior to 1922 as part of the aircraft designation to indicate various versions of a basic model. A few of these early suffix letters used by manufacturers are shown below:

A	Miscellaneous modification
B	Miscellaneous modification
D	Miscellaneous modification
G	Gun on a normally unarmed aircraft
GF	Armed Observation version
H	Aircraft fitted with the American-built Hispano-Suiza (Hisso) engine.
L	Aircraft fitted with the Liberty engine
SF	Aircraft converted to single-seat fighter

11 CARRIER AIRCRAFT ASSIGNMENT

August 1926

AIRCRAFT SQUADRONS, BATTLE FLEET

CV-1 LANGLEY

VF-1	1 F6C-1, 17 TS-1, 1 VE-9
VF-2	3 F6C-1, 3 F6c-2, 4 FB-1, 1 FB-3, 3 VE-7S
VT-2	4 SC-1, 12 SC-2
VJ-1	2 H-16, 2 MO-1, 1 NB-1, 1 PN-7

August 1927

AIRCRAFT SQUADRONS, BATTLE FLEET

CV-1 LANGLEY

VF-1B	12 FB-5
VF-2B	14 FU-1, 1 F6C-2, 1 UO-1
VF-6B *	13 FB-5, 1 DT-2
* formerly VF-2	
VT-2B	16 T3M-2, 4 SC-1, 9 SC-2, 1 T2D-1
VJ-1B	1 F-5L, 5 FU-1, 2 H-16, 1 NB-1, 1 PN-7, 1 PN-10, 1 OL-1, 1 TB-1, 3 UO-1C

AIRCRAFT SQUADRONS, SCOUTING FLEET

CV-2 LEXINGTON

VT-1S	16 T3M-2
VJ-2S	2 UO-1C

LEXINGTON and SARATOGA DETACHMENTS at Hampton Roads

VF-5S	18 F6C-3

August 1928

AIRCRAFT SQUADRONS, BATTLE FLEET

CV-3 SARATOGA

VF-1B	15 F2B-1
VB-2B *	16 F2B-1, 1 UO-1C
* formerly VF-6B	
VS-2B	no aircraft assigned
VT-2B	14 T3M-2, 10 T4M-1

CV-2 LEXINGTON

VF-3B	12 FB-5, 1 UO-1
VB-1B *	2 F6C-2, 17 F6C-3, 1 UO-1C
* formerly VF-5S	
VS-3B	no aircraft assigned
VT-1B	9 T3M-2, 10 T4M-1

CV-1 LANGLEY

VF-2B	4 F3B-1, 8 FU-1, 1 FU-2, 1 UO-5
VS-1B	6 O2U-1, 2 UO-1C, 1 UO-5

April 1929

AIRCRAFT SQUADRONS, BATTLE FLEET

CV-3 SARATOGA

VF-1B	19 F2B-1, 1 FU-2
VB-2B	18 F3B-1, 1 FU-2
VS-2B	12 O2U-2
VT-2B	27 T4M-1

CV-2 LEXINGTON

VF-3B	12 F3B-1, 1 FU-2
VB-1B	15 F3B-1
VS-3B	12 O2U-2
VT-1B	28 T4M-1

CV-1 LANGLEY

VF-2B	18 F3B-1, 1 FU-2
VS-1B	6 O2U-2

June 1930

AIRCRAFT SQUADRONS, BATTLE FORCE

CV-1 LANGLEY

VF-2B	8 F2B-1, 8 F6C-4
VS-1B	7 O2U-4, 1 O2U-2

CV-3 SARATOGA

VF-1B	no aircraft assigned
VF-6B	20 F3B-1
VS-2B	14 O2U-2
VT-2B	9 T4M-1, 3 TG-1
VJ-1B Detachment	5 OL-8

CV-2 LEXINGTON

VF-3B	20 F3B-1
VF-5B *	18 F4B-1
* formerly VB-1B	
VS-3B	13 O2U-4
VT-1B	19 T4M-1

June 1931

AIRCRAFT, BATTLE FORCE

CV-3 SARATOGA

VF-1B	18 F8C-4
VF-6B	18 F4B-2, 1 F3B-1, 1 O2C-2, 1 FU-2
VS-2B	12 O2U-2, 1 O2U-3
VT-2B	14 T4M-1, 7 TG-1

CV-2 LEXINGTON

VF-2B	18 F3B-1, 1 FU-2

VF-5B 11 F4B-1, 8 F4B-2, 1 O2C-2
VS-3B 4 O2U-2, 6 O2U-4
VT-1B 17 T4M-1

AIRCRAFT, SCOUTING FORCE

CV-1 LANGLEY

VS-1S 7 O2U-4, 1 O2U-3
VF-3S 18 F3B-1

October 1932

AIRCRAFT, BATTLE FORCE

CV-3 SARATOGA

VF-6B 13 F3B-1, 9 F4B-4, 1 SU-1
VF-1B 19 F4B-3, 1 SU-1
VS-2B 15 SU-1, 2 OL-9, 1 XO2L-1
VT-2B 21 TG-2

CV-1 LANGLEY

VF-3B 6 F3B-1, 11 F4B-4, 1 SU-1
VS-1B 2 O2U-2, 4 O2U-3, 6 O2U-4

AIRCRAFT, SCOUTING FORCE

CV-2 LEXINGTON

VF-2S 10 F4B-1, 8 F4B-2, 1 SU-2, 1 O2U-1
VF-5S 19 F4B-2, 1 SU-2
VS-3S 13 SU-2, 1 OL-8
VT-1S 15 BM-1

June 1933

AIRCRAFT, BATTLE FORCE

CV-3 SARATOGA

VF-1B 14 F11C-2, 19 F4B-3, 1 SU-3
VF-6B 20 F4B-4, 1 SU-3
VS-2B 5 SU-2, 2 SU-3, 5 SU-1, 2 OL-9

	VT-2B	4 TG-1, 18 TG-2
CV-2 LEXINGTON		
	VF-2B	14 F4B-2, 3 F4B-1, 1 SU-2
	VF-5B	18 F4B-2, 1 XFF-1, 1 SU-2
	VS-3B	3 SU-2, 8 SU-3
	VT-1B	6 BM-1, 12 BM-2
CV-1 LANGLEY		
	VF-3B	18 F4B-4, 1 SU-1
	VS-1B	7 SU-1, 4 O2U-3, 1 O2U-4

June 1934

AIRCRAFT, BATTLE FORCE

CV-3 SARATOGA		
	VF-1B	18 BFC-2, 1 SF-1
	VF-6B	20 F4B-4, 1 SF-1, 1 SU-3
	VS-2B	7 SU-1, 4 SU-2, 3 SU-3
	VT-2B	18 TG-2
CV-2 LEXINGTON		
	VF-2B	20 F4B-2, 1 SF-1, 1 SU-2
	VF-5B	19 FF-1
	VS-3B	8 SU-2, 6 SU-3
	VT-1B	12 BM-1, 6 BM-2

June 1935

AIRCRAFT, BATTLE FORCE

CV-3 SARATOGA		
	VB-2B *	20 BFC-2, 1 O2U-2, 1 SF-1
	* formerly VF-1B	
	VF-6B	17 F4B-4, 1 O2U-4, 1 SF-1
	VS-2B	9 SU-2, 9 SU-3
	VT-2B	18 TG-2

CV-2 LEXINGTON

VF-2B	21 F2F-1, 3 F4B-4, 1 O2U-4, 1 SU-1
VF-5B	21 FF-1, 1 O2U-4, 1 SF-1
VS-3B	18 SF-1
VB-1B *	7 BM-1, 12 BM-2

* formerly VT-1B

CV-4 RANGER

VF-3B	18 F2F-1, 19 F4B-4, 1 O3U-1, 1 SU-2, 1 SF-1
VB-3B	18 BG-1
VB-5B	15 BF2C-1, 1 O2U-2, 1 SF-1
VS-1B	20 SU-4

June 1936

AIRCRAFT, BATTLE FORCE

CV-3 SARATOGA

VB-2B	18 BFC-2, 1 SBU-1
VF-6B	16 F3F-1, 13 F4B-4, 1 SBU-1, 1 O3U-1
VS-2B	19 SBU-1
VT-2B	18 TG-2

CV-2 LEXINGTON

VB-3B	17 BG-1
VB-5B	10 F4B-4, 1 SBU-1, 1 SU-4
VF-2B	17 F2F-1, 1 SBU-1, 1 O3U-1
VS-3B	18 SBU-1

CV-4 RANGER

VB-1B	12 BM-1, 8 BM-2
VF-3B	17 F2F-1, 1 SBU-1, 1 O3U-1
VF-5B	18 F3F-1, 1 SBU-1, 1 O3U-1
VS-1B	19 SBU-1

CV-1 LANGLEY

VF-1B	18 F4B-4, 1 SU-4
VS-4B	11 SU-4

June 1937

AIRCRAFT, BATTLE FORCE

CARRIER DIVISION ONE

CV-3 SARATOGA

VF-6B	19 F3F-1, 1 SBU-1, 1 O3U-1
VB-2B	18 BFC-2, 1 SBU-1, 2 O3U-1
VS-2B	18 SBU-1
VS-4B	11 SU-4
VT-2B	17 TG-2

CV-2 LEXINGTON

VF-2B	18 F2F-1, 1 SBU-1, 1 O3U-1
VB-3B	18 BG-1
VS-3B	18 SBU-1
VB-5B	19 F4B-4, 1 SBU-1, 2 O3U-1

CARRIER DIVISION TWO

CV-5 YORKTOWN

VF-7B *	6 F2F-1
* formerly VF-3B	
VB-7B	1 SU-3
VS-7B	2 SU-2
VT-7B	1 SU-2, 1 SU-3

CV-6 ENTERPRISE

VF-8B *	16 F4B-4
* formerly VF-1B	
VS-8B	1 SU-4

CV-4 RANGER

VB-1B	8 BM-1, 10 BM-2
VF-5B	17 F3F-1, 1 SBU-1
VS-1B	18 SBU-1

June 1938

AIRCRAFT, BATTLE FORCE

CARRIER DIVISION ONE

CV-3 SARATOGA

VB-3	21 SB2U-1, 1 XSBA-1, 1 SBC-3, 1 N2Y-1
VF-3	16 F3F-1, 1 SBC-3, 2 O3U-3
VS-3	21 SBC-3
VT-3	21 TBD-1

CV-2 LEXINGTON

VB-2	22 SB2U-1
VF-2	20 F2F-1, 1 SBU-1, 2 O3U-3
VS-2	20 SBU-1
VT-2	21 TBD-1

CV-4 RANGER

VB-4	19 BG-1
VF-4	19 F3F-1, 1 SBU-1, 2 O3U-3
VS-41	22 SBU-1
VS-42	19 SBU-1

CARRIER DIVISION TWO

CV-5 YORKTOWN

VB-5	17 BT-1, 1 O3U-3
VF-5	19 F2F-1, 1 O3U-3
VS-5	10 SBC-3
VT-5	20 TBD-1

CV-6 ENTERPRISE

VB-6	13 BT-1
VF-6	20 F3F-2, 2 O3U-3, 1 SBC-3
VS-6	20 SBC-3
VT-6	20 TBD-1

June 1939

AIRCRAFT, BATTLE FORCE

CARRIER DIVISION ONE

CV-3 SARATOGA

VB-4	18 SB2U-2, 1 N2Y-1
VF-2	17 F2F-1, 1 SB2U-2, 2 SU-2, 1 SU-1, 1 O3U-3
VS-2	1 SBU-1, 1 SBC-4
VT-3	16 TBD-1, 1 SB2U-2

CV-2 LEXINGTON

VB-2	17 SB2U-1, 2 SB2U-2
VF-3	18 F3F-1, 1 SB2U-1, 2 SU-3
VS-3	18 SBC-3
VT-2	18 TBD-1

CV-4 RANGER

VB-3	2 SB2U-1
VF-4	20 F3F-1, 1 SBU-1, 2 SU-3
VS-41	18 SBU-1
VS-42	18 SBU-1

CARRIER DIVISION TWO

CV-5 YORKTOWN

VB-5	18 BT-1
VF-5	18 F3F-3, 1 F2F-1, 1 SU-3
VS-5	18 SBC-3
VT-5	18 TBD-1

CV-6 ENTERPRISE

VB-6	18 BT-1
VF-6	16 F3F-2, 2 SBC-3, 2 SU-2
VS-6	17 SBC-3
VT-6	17 TBD-1

June 1940

AIRCRAFT, BATTLE FORCE

CARRIER DIVISION ONE

CV-3 SARATOGA

VB-3	19 SB2U-2
VF-3	10 F2A-1, 9 F3F-1, 1 SB2U-2
VS-3	21 SBC-3
VT-3	22 TBD-1

CV-2 LEXINGTON

VB-2	15 SB2U-1, 4 SB2U-2
VF-2	21 F2F-1, 1 SB2U-1
VS-2	21 SBC-4
VT-2	21 TBD-1

CARRIER DIVISION TWO

CV-5 YORKTOWN

VB-5	22 BT-1, 1 SBC-3
VF-5	22 F3F-3, 1 SBC-3
VS-5	19 SBC-3
VT-5	23 TBD-1

CV-6 ENTERPRISE

VB-6	17 BT-1, 1 SBC-3, 2 SU-3
VF-6	17 F3F-2, 1 SBC-4, 2 SU-2
VS-6	17 SBC-3, 2 SU-2
VT-6	18 TBD-1, 1 SU-2, 1 SU-3

ATLANTIC SQUADRON

CV-4 RANGER

VB-4	13 SB2U-1, 3 SB2U-2, 3 SBC-3
VF-4	9 F3F-1, 2 F3F-3, 2 SU-3, 1 SBC-3
VS-41	16 SBU-1
VS-42	18 SBU-1

CV-7 WASP

VB-7	17 BG-1

	VF-7	15 F3F-1, 1 F2F-1, 1 SB2U-2, 2 SU-2
	VS-71	17 SBU-1, 1 J2F-1, 1 NY-3
	VS-72	19 SB2U-2

December 1941

CV-2 LEXINGTON		
	VF-2	19 F2A-3
	VS-2	2 SBD-2, 14 SBD-3
	VB-2	16 SBD-2
	VT-2	12 TBD-1
CV-3 SARATOGA		
	VS-3	22 SBD-3
	VF-3	2 F4F-3A, 8 F4F-3
	VB-3	21 SBD-3
	VT-3	12 TBD-1
CV-4 RANGER		
	VS-42	9 SB2U-1, 6 SB2U-2
	VS-41	8 SB2U-1, 2 SB2U-2
	VF-5	18 F4F-3, 1 F4F-3A
	VF-41	17 F4F-3, 2 SNJ-3
	VT-4	3 TBD-1
CV-5 YORKTOWN		
	VF-42	18 F4F-3
	VS-5	19 SBD-3, 2 SNJ-3
	VB-5	19 SBD-3
	VT-5	14 TBD-1
CV-6 ENTERPRISE		
	VS-6	10 SBD-2, 8 SBD-3
	VF-6	17 F4F-3A, 1 F4F-3
	VB-6	18 SBD-2
	VT-6	2 SNJ-3, 18 TBD-1
CV-7 WASP		
	VS-71	4 SB2U-1, 13 SB2U-2, 2 TBD-1

VF-71	18 F4F-3
VF-72	17 F4F-3, 2 SB2U-2, 1 SNJ-3
VS-72	18 SB2U-3

CV-8 HORNET

VS-8	20 SBC-4
VF-8	19 F4F-3, 2 F4F-3A, 2 SNJ-3
VB-8	19 SBC-4
VT-8	8 TBD-1, 7 SBN-1

AVG-1 LONG ISLAND

VS-201	12 SOC-3A, 1 SOC-1A, 7 F2A-3

1942

'DOOLITTLE RAID' on Tokyo -- 18 April

TASK FORCE 16

CV-6 ENTERPRISE

VB-6	18 SBD-2, -3
VF-6	27 F4F-3, -4
VT-6	18 TBD-1
VB-3 *	18 SBD-2, -3

* from SARATOGA to replace VS-6 which was training at Pearl Harbor

CV-8 HORNET

Strike Group * 16 B-25

* HORNET's own aircraft were packed into the hanger deck unable to be used until after the B-25s were launched

BATTLE OF CORAL SEA -- 7-8 May

TASK FORCE 17

TASK GROUP 17.5 CARRIER GROUP

CV-5 YORKTOWN (damaged on 8 May by Japanese dive-bomber)

VF-42	20 F4F-3
VB-5	19 SBD-2, -3
VS-5	19 SBD-2, -3
VT-5	13 TBD-1

CV-2 LEXINGTON (heavily damaged by Japanese dive and torpedo-bombers and had to be sunk on 8 May)

VF-2	22 F4F-3
VB-2	18 SBD-2, -3
VS-2	18 SBD-2, -3
VT-2	12 TBD-1

BATTLE OF MIDWAY -- 4-6 June

TASK FORCE 16 CARRIER STRIKE FORCE

TASK GROUP 16.5 CARRIER GROUP

CV-6 ENTERPRISE

VF-6	27 F4F-4
VB-6	19 SBD-2, -3
VS-6	19 SBD-2, -3
VT-6	14 TBD-1

CV-8 HORNET

VF-8	27 F4F-4
VB-8	19 SBD-2, -3
VS-8	18 SBD-1, -2, -3
VT-8	15 TBD-1

TASK FORCE 17 CARRIER STRIKE FORCE

TASK GROUP 17.5 CARRIER GROUP

CV-5 YORKTOWN (heavily damaged by dive and torpedo-bombers on 4 June and sunk by Japanese sub as she retired from the battle on 7 June)

VF-3	25 F4F-4
VB-3	18 SBD-3
VS-5	19 SBD-3
VT-3	13 TBD-1

INVASION OF MOROCCO -- 8-12 October

TASK FORCE 34 WESTERN NAVAL TASK FORCE

TASK GROUP 34.8 NORTHERN ATTACK GROUP

(AIR GROUP)

ACV-26 SANGAMON

VGS-26	9 TBF-1, 9 SBD-3
VGF-26	12 F4F-4

ACV-28 CHENANGO

76 Army P-40F for bases in Casablanca when captured

TASK GROUP 34.2 AIR GROUP

CV-4 RANGER

VF-9	27 F4F-4
VF-41	27 F4F-4
VS-41	18 SBD-3

ACV-27 SUWANNEE

VGF-27	11 F4F-4
VGS-27	9 TBF-1
VGF-28	12 F4F-4
VGS-30	6 F4F-4

TASK GROUP 34.10 SOUTHERN ATTACK GROUP

(AIR GROUP)

ACV-29 SANTEE

VGF-29	14 F4F-4
VGS-29	8 TBF-1, 9 SBD

1943

ESCORT CARRIERS IN THE ATLANTIC -- August

CVE-9 BOGUE

VC-9	12 F4F-4, 12 TBF-1

CVE-11 CARD

VC-1 6 F4F-4, 11 TBF-1

CVE-13 CORE

VC-13 6 F4F-1, 12 TBF-1

CVE-29 SANTEE

VC-29 13 TBF-1, 9 SBD-5

VF-29 12 F4F-4

GILBERT ISLANDS OPERATION -- 18-29 Nov

FIFTH FLEET

TASK FORCE 52 NORTHERN ATTACK FORCE

TASK GROUP 52.3 AIR SUPPORT GROUP

CVE-56 LISCOME BAY (sunk by Japanese sub on 24 November)

16 FM-1, 12 TBM-1

CVE-57 CORAL SEA

16 FM-1, 12 TBF-1

CVE-58 CORREGIDOR

16 FM-1, 12 TBF-1

TASK FORCE 53 SOUTHERN ATTACK FORCE

TASK GROUP 53.6 AIR SUPPORT GROUP

CVE-26 SANGAMON

9 TBF-1, 9 SBD-5, 12 F6F-3

CVE-27 SUWANNEE

9 SBD-5, 9 TBF-1, 12 F6F-3

CVE-28 CHENANGO

9 SBD-5, 9 TBF-1, 12 F6F-3

CVE-20 BARNES

VF-1 * 22 F6F-3

CVE-16 NASSAU

VF-1 * 22 F6F-3

* to form the air garrison on Tarawa after it was captured

TASK FORCE 50 FAST CARRIER FORCE

TASK GROUP 50.1 CARRIER INTERCEPTOR GROUP

CV-10 YORKTOWN

VF-5	36 F6F-3
VB-5	36 SBD-5
VT-5	18 TBF-1

CV-16 LEXINGTON

VB-16	36 SBD-5
VF-16	36 F6F-3
VT-16	18 TBF-1

CVL-25 COWPENS

VF-25	24 F6F-3
VF-6	12 F6F-3
VC-25	10 TBF-1

19 November - strikes on Jaluit and Mili
20-23 November - kept nearest Japanese airfields in the Marshall Islands pounded down and covered landings on Makin and Tarawa

TASK GROUP 50.2 NORTHERN CARRIER GROUP

CV-6 ENTERPRISE

VB-6	36 SBD-5
VT-6	18 TBF-1
VF-2	36 F6F-3

CVL-24 BELLEAU WOOD

VF-24	26 F6F-3
VF-6	12 F6F-3
VC-22B	9 TBF-1

CVL-26 MONTEREY

VF-30	24 F6F-3
VC-30	9 TBF-1

19-20 November - strikes on Makin and then assisted TF.52 as a striking group

TASK GROUP 50.3 SOUTHERN CARRIER GROUP

CV-9 ESSEX

VF-9	36 F6F-3
VB-9	36 SBD-5
VT-9	18 TBF-1

CV-17 BUNKER HILL

VB-17	32 SB2C-1
VF-18	36 F6F-3

CVL-22 INDEPENDENCE (heavily damaged on 20 November by torpedo bombers)

VF-22	16 F6F-3
VF-6	12 F6F-3
VC-22	9 TBF-1

18-20 November - strikes on Tarawa and Betio, then assisted TF.53 as a support group

TASK GROUP 50.4 RELIEF CARRIER GROUP

CV-3 SARATOGA

VF-12	37 F6F-3
VB-12	24 SBD-5
VT-12	18 TBF-1

CVL-23 PRINCETON

VF-23	24 F6F-3
VT-23	9 TBF-1

19 November - neutralized Nauru and then provided air cover for Makin and Tarawa air garrisons until they were operational

KWAJALEIN AND WOTJE STRIKES -- 4 December

TASK FORCE 50 FAST CARRIER FORCE

TASK GROUP 50.1

CV-10 YORKTOWN

VF-5	38 F6F-3
VB-5	28 SBD-5
VT-5	18 TBF-1

CV-16 LEXINGTON (damaged by torpedo bomber on 4 December)

VF-16	38 F6F-3
VB-16	28 SBD-5
VT-16	18 TBF-1

CVL-25 COWPENS

VF-25 VF-6	21 F6F-3
VC-25	9 TBF-1

TASK GROUP 50.3

CV-9 ESSEX

VF-9	38 F6F-3
VB-9	28 SBD-5
VT-9	18 TBF-1

CV-6 ENTERPRISE

VF-2	34 F6F-3
VB-6	20 SBD-5
VT-6	17 TBF-1

CVL-24 BELLEAU WOOD

VF-24 VF-6	24 F6F-3
VC-22B	9 TBF-1

1944

MARSHALL ISLANDS OPERATIONS -- 29 Jan - 28 Feb

FIFTH FLEET

TF 52 SOUTHERN ATTACK FORCE

TG 52.9 CARRIER SUPPORT GROUP

CVE-61 MANILA BAY

	VC-7	9 TBF-1, 3 TBF-1C, 16 FM-2

CVE-57 CORAL SEA

	VC-33	6 TBF-1C, 4 TBF-1, 2 TBM-1, 5 FM-1, 9 F4F-4

CVE-58 CORREGIDOR

	VC-44	8 TBM-1, 3 TBF-1C, 3 F4F-4, 6 FM-1

(provided air cover and anti-sub patrols)

TF 53 NORTHERN ATTACK FORCE

TG 53.6 CARRIER GROUP

CVE-26 SANGAMON

	VF-37	12 F6F-3
	VC-37	9 SBD-5, 4 TBM-1C, 6 TBF-1

CVE-27 SUWANNEE

	VF-60	12 F6F-3
	VC-60	9 SBD-5, 2 TBF-1, 7 TBM-1C

CVE-28 CHENANGO

	VF-35	12 F6F-3
	VC-35	4 TBF-1, 5 TBM-1C, 9 SBD-5

(provided air cover and anti-sub patrols
17 February - strikes on Eniwetok and then stood by as a support group until 28 February)

TG 51.2 MAJURO ATTACK GROUP

CVE-16 NASSAU

	VC-66	14 FM-1, 5 TBM-1C, 4 TBM-1

CVE-62 NATOMA BAY

	VC-63	6 TBM-1, 6 TBM-1C, 12 FM-1

TF 58 FAST CARRIER FORCE

TG 58.1 CARRIER TASK GROUP

CV-6 ENTERPRISE

	VB-10	30 SBD-5
	VF-10	32 F6F-3
	VT-10	16 TBF-1C

CV-10 YORKTOWN

	VB-5	36 SBD-5, 1 F6F-3
	VF-5	36 F6F-3
	VT-5	16 TBF-1, 2 TBF-1C

CVL-24 BELLEAU WOOD

	VF-24	22 F6F-3
	VT-24	6 TBF-1, 2 TBF-1C

29 January - strikes on Maloelap
30 January - strikes on Kwajalein
31 January - covered landings on Kwajalein and then pro-vided ground support until 3 February

TG 58.2 CARRIER TASK GROUP

CV-9 ESSEX

	VB-9	34 SBD-5, 1 F6F-3
	VF-9	35 F6F-3
	VT-9	15 TBF-1, 2 TBF-1C, 2 TBM-1C

CV-11 INTREPID

	VB-6	36 SBD-5
	VF-6	37 F6F-3
	VT-6	7 TBF-1, 7 TBF-1C, 5 TBM-1C

CVL-28 CABOT

	VF-31	24 F6F-3
	VT-31	8 TBM-1C

29-30 January - strikes on Roi
31 January - covered landings on Roi and then provided ground support until 3 February

TG 58.3 CARRIER TASK GROUP

CV-17 BUNKER HILL

	VB-17	31 SB2C-1, 1 F6F-3
	VF-18	37 F6F-3
	VT-17	18 TBF-1C, 1 TBF-1, 1 TBM-1C

CVL-26 MONTEREY

	VF-30	25 F6F-3
	VT-30	3 TBF-1C, 2 TBF-1, 4 TBM-1C

CVL-25 COWPENS

	VF-25	22 F6F-3
	VT-25	9 TBF-1C

29 January - strikes on Kwajalein
30 January-1 February - strikes on Eniwetok

TG 58.4 CARRIER TASK GROUP

CV-3 SARATOGA

	VB-12	24 SBD-5
	VF-12	36 F6F-3
	VT-12	18 TBM-1C

CVL-23 PRINCETON

	VF-23	24 F6F-3
	VT-23	8 TBF-1, 1 TBM-1C

CVL-27 LANGLEY

	VF-32	22 F6F-3
	VT-32	9 TBF-1C

29 January - strikes on Wotje
30-31 January - strikes on Wotje and Maloelap
3-6, 11, 13 and 17 February - strikes on Eniwetok and then stood by as a support group until 28 February

STRIKES ON TRUK -- 17-18 Feb

TF 58 FAST CARRIER FORCE

TG 58.1 CARRIER GROUP

CV-6 ENTERPRISE

same as Marshalls Operation plus:
VF(N)-101(Detachment A) 4 F4U-2

CV-10 YORKTOWN

same as Marshalls Operation plus:
VF(N)-76(Detachment B) 4 F6F-3N

CVL-24 BELLEAU WOOD

same as Marshalls Operation

TG 58.2 CARRIER GROUP

CV-9 ESSEX

same as Marshalls Operation

CV-11 INTREPID (damaged by torpedo bomber on 17 Feb)

same as Marshalls Operation plus:
VF(N)-101(Detachment B) 4 F4U-2

CVL-28 CABOT

same as Marshalls Operation

TG 58.3 CARRIER GROUP

CV-17 BUNKER HILL

same as Marshalls Operation plus:
VF(N)-76(Detachment A) 4 F6F-3N

CVL-26 MONTEREY

same as Marshalls Operation

CVL-25 COWPENS

same as Marshalls Operation

ESCORT CARRIERS IN THE ATLANTIC -- May

CVE-9 BOGUE

VC-69 9 FM-2, 12 TBM-1C after 5 May

CVE-21 BLOCK ISLAND (sunk by German sub on 29 May)
VC-58 9 FM-1, 12 TBF-1 till 15 May
VC-55 9 FM-1, 12 TBF-1 after 23 May

CVE-25 CROATAN
VC-42 9 FM-2, 11 TBM-1C till 12 May

CVE-60 GUADALCANAL
VC-8 12 TBM-1C, 9 FM-2 after 14 May

CVE-67 SOLOMONS
VC-9 9 FM-2, 6 TBM-1, 6 TBF-1 after 30 May

INVASION OF LEYTE -- 17-25 OCTOBER

SEVENTH FLEET

TF 77 CENTRAL PHILIPPINES ATTACK FORCE

TG 77.4 ESCORT CARRIER GROUP

77.4.1 "TAFFY 1"

CVE-26 SANGAMON (damaged by bombers on 19 October)
(damaged by Kamikaze on 25 October)
Air Group 37 12 F6F-3, 5 F6F-5, 9 TBM-1C

CVE-27 SUWANNEE (damaged by Kamikaze on 25 October)
(damaged by bombers and Kamikaze on 26 Oct.)
Air Group 60 22 F6F-3, 9 TBM-1C

CVE-28 CHENANGO
Air Group 35 22 F6F-3, 9 TBM-1C

CVE-29 SANTEE (damaged by submarine and Kamikaze on 25 Oct.)
Air Group 26 24 FM-2, 6 TBF-1C, 3 TBM-1C

CVE-82 SAGINAW BAY
VC-78 15 FM-2, 12 TBM-1C

CVE-80 PETROF BAY
VC-76 16 FM-2, 10 TBM-1C

77.4.2 "TAFFY 2"

CVE-62 NATOMA BAY
VC-81 16 FM-2, 12 TBM-1C

CVE-61 MANILA BAY
VC-80 16 FM-2, 12 TBM-1C

CVE-77 MARCUS ISLAND
VC-21 12 FM-2, 11 TBM-1C

CVE-76 KADASHAN BAY
VC-20 15 FM-2, 11 TBM-1C

CVE-78 SAVO ISLAND
VC-27 16 FM-2, 12 TBM-1C

CVE-79 OMMANEY BAY
VC-75 16 FM-2, 11 TBM-1C

77.4.3 "TAFFY 3"

CVE-70 FANSHAW BAY (damaged by gunfire on 25 October)
VC-68 16 FM-2, 12 TBM-1C

CVE-63 ST. LO (ex-MIDWAY) (sunk by Kamikaze on 25 October)
VC-65 17 FM-2, 12 TBM-1C

CVE-66 WHITE PLAINS (damaged by gunfire and Kamikaze on 25 Oct.)
VC-4 16 FM-2, 12 TBM-1C

CVE-68 KALININ BAY (damaged by gunfire and Kamikaze on 25 Oct.)
VC-3 16 FM-2, 1 TBF-1C, 11 TBM-1C

CVE-71 KITKUN BAY (damaged by Kamikaze on 25 October)
VC-5 14 FM-2, 12 TBM-1C

CVE-73 GAMBIER BAY (sunk by gunfire on 25 October)
VC-10 18 FM-2, 12 TBM-1C

(TG 77.4)

Provided anti-sub patrols off Leyte Gulf, Combat Air Patrol for the convoy, and air cover for the landings.

THIRD FLEET

TF 38 FAST CARRIER FORCE

TG 38.1 TASK GROUP 1

CV-18 WASP

Air Group 14

VF-14	29 F6F-3, 3 F6F-3N, 2 F6F-3P 7 F6F-5, 1 F6F-5N
VB-14	3 F6F-3, 7 F6F-5, 25 SB2C-3
VT-14	5 TBF-1C, 1 TBD-1D, 11 TBM-1C, 1 TBM-1D

CV-12 HORNET

Air Group 11

VF-11	11 F6F-3, 2 F6F-3N, 1 F6F-3P, 21 F6F-5, 2 F6F-5N, 3 F6F-5P
VB-11	25 SB2C-3
VT-11	1 TBF-1C, 17 TBM-1C

CVL-26 MONTEREY

Air Group 28

VF-28	21 F6F-5, 2 F6F-5P
VT-28	9 TBM-1C

CVL-25 COWPENS (to TG 30.3 on 14 October)

Air Group 22

VF-22	25 F6F-5, 1 F6F-5P
VT-22	9 TBM-1C

TG 38.2 TASK GROUP 2

CV-11 INTREPID (damaged by Kamikaze on 29 October)

Air Group 18

VF-18	5 F6F-3N, 35 F6F-5, 3 F6F-5P
VB-18	28 SB2C-3
VT-18	18 TBM-1C

CV-19 HANCOCK (damaged by bombers on 14 October)
(to TG 38.1 22-26 October)

Air Group 7

VF-7	37 F6F-5, 4 F6F-5N
VB-7	30 SB2C-3, 12 SB2C-3E
VT-7	18 TBM-1C

CV-17 BUNKER HILL (detached on 23 October)

Air Group 8

VF-8	27 F6F-3, 13 F6F-5, 4 F6F-3N, 4 F6F-5N
VB-8	17 SB2C-1C, 3 SBF-1, 4 SBW-1
VT-8	17 TBM-1C, 2 TBM-1D

CVL-28 CABOT (to TG 30.3 on 14 October)

Air Group 29

VF-29	3 F6F-3, 18 F6F-5
VT-29	1 TBF-1C, 8 TBM-1C

CVL-22 INDEPENDENCE

Night Air Group 41

VFN-41	3 F6F-2, 2 F6F-5, 14 F6F-5N
VTN-41	8 TBM-1D

TG 38.3 TASK GROUP 3

CV-9 ESSEX

Air Group 15

VF-15	22 F6F-3, 3 F6F-3N, 2 F6F-3P, 22 F6F-5, 1 F6F-5N
VB-15	25 SB2C-3
VT-15	15 TBF-1C, 5 TBM-1C

CV-16 LEXINGTON

Air Group 19

VF-19	14 F6F-3, 2 F6F-3N, 1 F6F-3P, 21 F6F-5, 2 F6F-5N, 1 F6F-5P
VT-19	18 TBM-1C
VB-19	30 SB2C-3

CVL-23 PRINCETON (damaged by dive bombers and scuttled on 24 October)

Air Group 27

VF-27	18 F6F-3, 7 F6F-5
VT-27	9 TBM-1C

CVL-27 LANGLEY

Air Group 44

VF-44	19 F6F-3, 6 F6F-5
VT-44	9 TBM-1C

TG 38.4 TASK GROUP 4

CV-13 FRANKLIN (damaged by Kamikaze on 13 and 30 October)
(damaged by bombers on 15 October)

Air Group 13

VF-13	1 F6F-3, 3 F6F-3N, 29 F6F-5, 1 F6F-5N, 4 F6F-5P
VB-13	31 SB2C-3
VT-13	18 TBM-1C

CV-6 ENTERPRISE

Air Group 20

VF-20	4 F6F-3N, 35 F6F-5
VB-20	34 SB2C-3
VT-20	19 TBM-1C

CVL-30 SAN JACINTO

Air Group 51

VF-51	14 F6F-3, 5 F6F-5
VT-51	7 TBM-1C

CVL-24 BELLEAU WOOD (damaged by Kamikaze on 30 October)

Air Group 21

VF-21	24 F6F-5, 1 F6F-5P
VT-21	9 TBM-1C

(TF 38)

10-13 Oct. strikes on Okinawa, Formosa, and Northern Leyte.
16-20 Oct. strikes on Leyte, Cedu, Negros and support the landings on Leyte.
On and after 21 Oct. strategic support of Leyte operations

by engaging all enemy naval and air forces threatening the area.

TG 30.8 AT SEA LOGISTICS GROUP

CVE-20	BARNES	CVE-85	SHIPLEY BAY
CVE-18	ALTAMAHA	CVE-87	STEAMER BAY
CVE-86	SITKOH BAY	CVE-74	NEHENTA BAY
CVE-88	CAPE ESPERANCE	CVE-83	SARGENT BAY
CVE-16	NASSAU	CVE-81	RUDYERD BAY
CVE-98	KWAJALEIN		

Carried replacement aircraft for Task Force 38 and 77

TG 30.3 "CRIPDIV"

CVL-25	COWPENS	CVL-28	CABOT

Formed 15 Oct. to tow U.S.S. HOUSTON and U.S.S. CANBERRA to Ulithi anchorage.

1945

CAPTURE OF OKINAWA -- 14 Mar - 30 June

FIFTH FLEET
TF 52 AMPHIBIOUS SUPPORT FORCE
TG 52.1 SUPPORT CARRIER GROUP
UNIT 1

CVE-93 MAKIN ISLAND
VC-84 16 FM-2, 11 TBM-3

CVE-70 FANSHAW BAY
VOC-2 24 FM-2, 6 TBM-3

CVE-94 LUNGA POINT
VC-85 18 FM-2, 11 TBM-3, 1 TBM-3P

CVE-26 SANGAMON (damaged by Kamikaze on 4 May)
Air Group 33 24 F6F, 6 TBM-3

CVE-62 NATOMA BAY (damaged by Kamikaze on 6 June)
VC-81 20 FM-2, 11 TBM-1C, 1 TBM-1CP

CVE-78 SAVO ISLAND
VC-91 20 FM-2, 11 TBM-1C, 4 TBM-3

CVE-57 ANZIO (ex-CORAL SEA)
VC-13 12 FM-2, 12 TBM-1C

UNIT 2

CVE-82 SAGINAW BAY
VC-88 20 FM-2, 12 TBM

CVE-83 SARGENT BAY
VC-83 16 FM-2, 12 TBM-1C

CVE-81 RUDYERD BAY
VC-96* 20 FM-2, 11 TBM-1C

* operated from CVE-84 SHAMROCK BAY during 23 May - 30 June

CVE-77 MARCUS ISLAND
VC-87 20 FM-2, 12 TBM-3

CVE-80 PETROF BAY
VC-93 16 FM-2, 12 TBM-3

CVE-72 TULAGI
VC-92 19 FM-2, 12 TBM-3

CVE-65 WAKE ISLAND (damaged by Kamikaze on 3 April)
VOC-1 26 FM-2, 6 TBM-3

UNIT 3

CVE-27 SUWANNEE (damaged by on-board explosion on 24 May)
Air Group 40 17 F6F, 10 TBM

CVE-28 CHENANGO (damaged by crash of friendly aircraft on 9 April)
Air Group 25 17 F6F-5, 1 F6F-5P, 12 TBM

CVE-29 SANTEE

Air Group 24 18 F6F, 12 TBM

CVE-87 STEAMER BAY (damaged by collision on 25 April)

(TG 52.1) VC-90 19 FM-2, 12 TBM-3

Provided daily strikes on Okinawa, spotted for naval gunfire, flew photo-reccon missions, and conducted anti-sub patrols. After 26 March provided Combat Air Patrols

SPECIAL ESCORT CARRIER GROUP*

CVE-97 HOLLANDIA

CVE-66 WHITE PLAINS

CVE-86 SITKOH BAY

CVE-23 BRETON

* arrived 4 April

Brought up Marine Air Groups 31 and 33, with 192 F4U and 30 F6F, to operate from airfields on Okinawa.

TF 58 FAST CARRIER FORCE

TG 58.1*

CV-12 HORNET (damaged by typhoon on 5 June)

Air Group 17

VF-17 61 F6F-5, 6 F6F-5P, 4 F6F-5N

VB-17 9 SB2C-3, 2 SB2C-4, 4 SBW

VT-17 15 TBM-3

CV-18 WASP (damaged by bombers on 19 March)

Air Group 86

VBF-86 36 F4U-1D

VF-86 28 F6F-5, 2 F6F-5E, 2 F6F-5P, 2 F6F-5N

VB-86 15 SB2C-4

VT-86 15 TBM-3

CV-20 BENNINGTON (damaged by typhoon on 5 June)

Air Group 82

VF-82 29 F6F-5, 2 F6F-5E, 2 F6F-5P, 4 F6F-5N

VB-82 15 SB2C-4E

VT-82 15 TBM-3

VMF-112   18 F4U-1D

VMF-123   17 F4U-1D

CVL-24 BELLEAU WOOD (damaged by typhoon on 5 June)

Air Group 30

VF-30   24 F6F-5, 1 F6F-5P

VT-30   8 TBM-3, 1 TBM-3P

CVL-30 SAN JACINTO (damaged by Kamikaze on 6 April)
(damaged by typhoon on 5 June)

Air Group 45

VF-45   24 F6F-5, 1 F6F-5P

VT-45   9 TBM-3

* after 7 April - HORNET, BENNINGTON, BELLEAU WOOD, SAN JACINTO

TG 58.2*

CV-6 ENTERPRISE (damaged by bombers on 18 March)
(damaged by U.S. gunfire on 20 March)
(damaged by Kamikaze on 11 April and 13 May)

Night Air Group 90

VFN-90   11 F6F-5E, 2 F6F-5P, 19 F6F-5N

VTN-90   21 TBM-3D

CV-13 FRANKLIN (damaged by bombers on 19 March)

Air Group 5

VF-5   2 F6F-5P, 4 F6F-5N, 2 FG-1D, 30 F4U-1D

VB-5   15 SB2C-4E

VT-5   15 TBM-3

CV-15 RANDOLPH

Air Group 12

VF-12   25 F6F-5, 2 F6F-5E, 2 F6F-5P, 4 F6F-5N

VBF-12   24 F6F-5

VB-12   15 SB2C-4E

VT-12   15 TBM-3

*after 7 April - RANDOLPH, ENTERPRISE, INDEPENDENCE

TG 58.3*

CV-9 ESSEX (damaged by U.S. gunfire on 19 March)
(damaged by aircraft crash on 27 March)
(damaged by dive bombers on 11 April)

Air Group 83

VF-83	26 F6F-5, 2 F6F-5E, 2 F6F-5P, 4 F6F-5N
VBF-83	36 F4U-1D
VB-83	15 SB2C-4
VT-83	15 TBM-3

CV-17 BUNKER HILL (damaged by Kamikaze on 11 May)

Air Group 84

VF-84	6 F6F-5P, 4 F6F-5N, 27 F4U-1D
VB-84	2 SB2C-4, 13 SB2C-4E
VT-84	15 TBM-3
VMF-221	18 F4U-1D
VMF-451	18 F4U-1D

CV-19 HANCOCK (damaged by Kamikaze on 7 April)

Air Group 6

VF-6	28 F6F-5, 2 F6F-5E, 2 F6F-5P, 4 F6F-5N
VBF-6	36 F6F-5
VB-6	5 SB2C-3 & 3E, 3 SB2C-4, 4 SBW-3 & 4E
VT-6	10 TBM-3

CVL-28 CABOT

Air Group 29

VF-29	25 F6F-5
VT-29	9 TBM-3

CVL-29 BATAAN (damaged by U.S. gunfire on 13 May)

Air Group 47

VF-47	23 F6F-5, 1 F6F-5P
VT-47	12 TBM-3

*after 7 April - ESSEX, BUNKER HILL, BATAAN; SHANGRI-LA joined at the end of April)

TG 58.4*

CV-10 YORKTOWN (damaged by bombers on 18 March)

Air Group 9

VF-9	40 F6F-5,5E,5P,and 5N
VBF-9	33 F6F-5
VB-9	15 SB2C-4
VT-9	6 TBM-3, 1 TBM-3P

after 27 June - Air Group 88

VF-88	30 F6F-5, 3 F6F-5P, 6 F6F-5N
VBF-88	37 FG-1D
VB-88	9 SB2C-4, 6 SB2C-4E
VT-88	14 TBM-3, 1 TBM-3E

CV-11 INTREPID (damaged by Kamikaze on 18 March and 16 April)
(damaged by U.S. gunfire on 18 March)

Air Group 10

VF-10	2 F6F-5P, 4 F6F-5N, 1 FG-1, 29 F4U-1D
VBF-10	36 F4U-1D
VB-10	15 SB2C-4E
VT-10	15 TBM-3

CVL-27 LANGLEY

Air Group 23

VF-23	3 F6F-3, 21 F6F-5, 1 F6F-5P
VT-23	9 TBM-1C

CVL-22 INDEPENDENCE

Air Group 46

VF-46	24 F6F-5, 1 F6F-5P
VT-46	8 TBM-3

*after 7 April - YORKTOWN, INTREPID, LANGLEY; ENTERPRISE joined on 10 April.

(TF 58)

Air support prior to landings. After landings provided Combat Air Patrol over amphibious forces, plus strikes on Okinawa and anti-sub patrols.

TG 50.8 LOGISTICS SUPPORT GROUP

CVE-84 SHAMROCK BAY

VC-94 18 FM-2, 12 TBM-3

CVE-91 MAKASSAR STRAIT

VC-97 14 FM-2, 12 TBM-3

(Provided Combat Air Patrol for the Support Group)

TU 50.8.4 CVE PLANE TRANSPORT UNIT

CVE-102 ATTU (damaged by typhoon on 5 June)

CVE-99 ADMIRALTY ISLANDS

CVE-100 BOUGAINVILLE (damaged by typhoon on 5 June)

CVE-92 WINDHAM BAY (damaged by typhoon on 5 June)

(Carried replacement aircraft for the fleet and for the air units to be based on Okinawa after the fields were taken.)

An additional 17 CVEs carried aircraft from the U.S. West Coast to forward bases for deployement to the ATTU, ADMIRALTY ISLANDS, BOUGAINVILLE, and WINDHAM BAY.

Author's comments:

The Aircraft Assignment Tables for the period 1926 - 1941 is an excerpt from the annual Fleet Aircraft Assignment Tables in "U.S. NAVY AIRCRAFT 1921 - 1941" and is re-printed by per--mission of the author -- William T. Larkins.

The Aircraft Assignment Tables for the period 1942 - 1945 are based on tables found in Samuel Eliot Morison's "HISTORY OF U.S. NAVAL OPERATIONS IN WORLD WAR II", volumes I - XV. This is further supplemented by material found in the text of "HISTORY OF U.S. NAVAL OPERATIONS IN WORLD WAR II", "THE FAST CARRIERS", ON THE SEAS and IN THE SKIES", and "GRUMMAN F6F-3/5 HELLCAT".

REFERENCES

AIR PROFILES, Volume 1 - 10
Doubleday Publication

AIRCRAFT CARRIER - THE MAJESTIC WEAPON
Donald MacIntyre

CARRIER FIGHTERS, Volume 1 & 2
Sentry Books Publication

CHRONOLOGY OF CAMOUFLAGE and MARKINGS FOR U.S. NAVAL COMBAT AIRCRAFT WORLD WAR II
R.M. Hill (I.P.M.S./U.S.A.)

DIVE and TORPEDO BOMBERS
Sentry Books Publication

GRUMMAN F6F-3/5 HELLCAT
Aircram Aviation Series

HISTORY OF U.S. NAVAL OPERATIONS IN WORLD WAR II, Volume I - XV
Samuel Eliot Morison

INCREDIBLE VICTORY
Walter Lord

MARKINGS OF THE ACES, Part 2, U.S. Navy, Book 1 & 2
Richard Hill

ON THE SEAS and IN THE SKIES
Theodore Roscoe

SHARKMOUTH 1916 - 1970, Volume 1 & 2
Aircam Aviation Series

THE FAST CARRIERS
Clark G. Reynolds

UNITED STATES NAVAL AVIATION 1910 - 1970 (NAVAIR 00-80P-1)
U.S. Government Printing Office

U.S. CAMOUFLAGE WORLD WAR II
Jay Frank Dial

U.S. NAVY AIRCRAFT 1921 - 1941
William T. Larkins

U.S. NAVY AIRCRAFT SINCE 1911
Gordon Swanborough and Peter M. Bowers

U.S. NAVY and MARINE FIGHTERS 1918 - 1962
Paul R. Matt and Bruce Robertson

U.S. NAVY MARKINGS WORLD WAR II, PACIFIC THEATER
Thomas E. Doll

Information was also obtained from various issues of the following periodicals:

AERO MODELLER
AIR CLASSICS
AIR PICTORIAL
AIRPLANE SPOTTER
AIR PROGRESS
AMERICAN AVIATION HISTORICAL SOCIETY JOURNAL
FLYING
FLYING REVIEW INTERNATIONAL
INTERNATIONAL PLASTIC MODELERS SOCIETY / U.S. QUARTERLY
NAVAL AVIATION NEWS
POPULAR AVIATION
SCALE MODELLER
U.S. NAVAL INSTITUTE PROCEEDINGS

Also used were various Aviation Circular Letters and painting instructions made available by William T. Larkins. These dealt with World War II color schemes and markings plus I.D. letters.